BRAIN ART AND NEUROSCIENCE

T0314391

The first of its kind, this book examines artistic representations of the brain after the rise of the contemporary neurosciences, examining the interplay of art and science and tackling some of the critical-cultural implications.

Weaving an MRI pattern onto a family quilt. Scanning the brain of a philosopher contemplating her own death and hanging it in a museum. Is this art or science or something in-between? What does it mean? How might we respond? In this ground-breaking new book, David R. Gruber explores the seductive and influential position of the neurosciences amid a growing interest in affect and materiality as manifest in artistic representations of the human brain. Contributing to debates surrounding the value of interdisciplinary engagement happening in the neuro-humanities, Gruber emphasizes the need for critical-cultural analysis within the field. Engaging with New Materialism and Affect Theory, the book provides a current and concrete example of the on-going shift away from constructivist lenses, arguing that the influence of relatively new neuroscience methods (EEG, MRI and fMRI) on the visual arts has not yet been fully realised. In fact, the very idea of a brain as it is seen and encountered today—or "The Brain," as Gruber calls it—remains in need of critical, wild and rebellious re-imagination.

Illuminating how artistic engagement with the brain is often sensual and suggestive even if rooted in objectivist impulses and tied to scientific realism, this book is ideal for scholars in Art, Media Studies, Sociology, and English departments, as well visual artists and anyone seriously engaging discourses of the brain.

David R. Gruber is an Assistant Professor in the Department of Media, Cognition and Communication at the University of Copenhagen. His research spans the rhetoric of science, body studies, and the public understanding of science. Much of his work focuses on the role of neuroscience in society and the reasons why brain findings can be so persuasive. He is co-editor of *The Routledge Handbook of Language and Science* with Lynda Olman and creator of the neuro-humanities website, Neurohuman.com.

BRAIN ART AND NEUROSCIENCE

Neurosensuality and Affective Realism

David R. Gruber

LONDON AND NEW YORK

First published 2020
by Routledge
2 Park Square, Milton Park, Abingdon, Oxon OX14 4RN

and by Routledge
52 Vanderbilt Avenue, New York, NY 10017

Routledge is an imprint of the Taylor & Francis Group, an informa business

British Library Cataloguing-in-Publication Data
A catalogue record for this book is available from the British Library

Library of Congress Cataloging-in-Publication Data
A catalog record has been requested for this book

ISBN: 978–0–367–89818–2 (hbk)
ISBN: 978–0–367–89819–9 (pbk)
ISBN: 978–1–003–02130–8 (ebk)

Typeset in Bembo
by Integra Software Services Pvt. Ltd.

CONTENTS

ACKNOWLEDGMENTS

Many people work together to bring a book into existence. I want to thank Vilija Stephens for leading the publication process at Routledge. Thanks also goes to Jason Kalin for reading an early draft, for being patient with my ramblings, and for providing much-needed guidance. Deep gratitude goes to Lindsey Gruber for her many long hours working through these ideas with me and for not being afraid to say that I am making no sense (true acts of love). I would also like to thank David Rieder for teaching me that making art is a way of building ideas unimaginable otherwise. All of these people, and more, have supported me and contributed to this book.

PREFACE

To read this book, you must pull your heart out of your chest and in its beating rhythm see a brain. You must watch the traffic and be blinded by the flow of neuronal pathways. Stare up at the night sky and see only bursts of oxygen inside of your skull. After this, you must look at a brain scan and focus your eyes deep into the pitch black lingering ominously behind the painted blue blobs of neural activity; there, absorb the immensity of human ignorance stirring in and around the impenetrable dark core of the amygdala.

You must scan your own brain to read this book. Transpose your darkest secrets onto the baseline BOLD variance.[1] Whisper your dreams in neuroscience discourses. Lay the body passively down. Allow the chill of the air-conditioning and the smell of cleaning fluid in the neuroscience lab to be but the passing of positively and negatively charged ions across neural membranes. Then march forward into the tunnel of light in the functional magnetic resonance imaging tube like an exhilarating journey to heaven. Emerge anew and plunge into the brain scan like a baptismal bath. But be prepare to encounter the unimaginable. Be prepared to dislodge all of your expectations.

Imagine now that you are walking into a movie theater. You sit down and realize, quite suddenly, that you have seen this film before. Then you see yourself on screen acting in the film. You feel every detail in your chest. The film is replaying your own memories. And then you know: the projection is you. The reel is your frontal lobe firing away. Your eyes are doing the flickering. You run outside to escape what feels like a nightmare. Gasping for breath, you look up at the sky. A million terrible thunderstorms roar across the horizon—it's all neurons firing. You are inside of your own brain. An apocalyptic dream, no, a revelation. You stand in a concert crescendo of Die Walküre materially enacted. You scream and hear your own voice echo back, but when you do, the sky glows a brilliant and hellish red. Then you scream for the fright of it.

You wake up in bed. The comfort of home. You grab your head and feel the contours of your skull. You turn to the mirror and breathe a sigh of relief when you see your face, red and sweating. You wipe your forehead, touch your chest, and take a deep breath. It feels good to have a body again. You turn to the window, throw open the curtains. Suddenly, a glowing 200-foot-tall neuron, whirling like a giant electric octopus arm, zaps you with a shuddering bolt. It obliterates your existence.

The book's argument

As a brain, we can toggle back and forth between the real and the imagined, a dream state and a state of Being. We can Be something definable and yet Become much more. We can sense the world but never know if it is an illusion, imagining only, perhaps, what keeps us alive. Yet, we feel that something lurks within the brain that it cannot or does not itself think. We then agree that we are prisoners of this brain, but in a twist of fate, we do not know if there are any guards nor how the walls are made. Neuronal structures seal us with a sobriety, yet we are drunk on creativity about what the brain might be able to do. We are not sure yet what it will mean for the human future to alter the brain. We do not so much want to escape it but to roam its careening hallways, cultivate its wild gardens, and expand its visionary vistas.

But the optimism can only go so far. Every time we forget a name or can't roll out of bed in the morning without the thought of a cup of coffee to motivate us, we feel a little unsure about making ourselves into a brain. We question any theatrical Captain Kirk brain of commander rationalism confidently standing up there to save us from ourselves. We double over with a sick stomach at the dead flat yet all-encompassing description of the dictionary: "*the brain is:* the organ of soft nervous tissue contained in the skull of vertebrates, functioning as the coordinating center of sensation and intellectual and nervous activity."[2] We feel overcome with the enormity yet ridiculous simplicity of it, "center of *sensation* and *intellectual* and *nervous* activity." We get nervous; we just bought a shitload of probiotics and need to run to the pharmacy to renew a prescription for Monopril; we freaked out at the thought of attending the family reunion with so many faces after twenty years; we hid in the bathroom for forty minutes during an international conference just to avoid social contact; we dialed the wrong number twice last week; we had a concussion in sixth grade and then another in college; we got food poisoning in Costa Rica and couldn't respond to work emails for three days; we have no idea who "Ida" from high school might be and why the hell she wants to be friends on Facebook.

In the simple dictionary description, we see the world in the head of a pin, which is sitting on our shoulders. We discover so much more than an organ there in that description and so much less than all that goes into the production of just one thought. This is what makes us feel queasy: extending the brain's controlling power to every sensation of reality at the same time as reducing every sensation of reality to the action of "soft nervous tissue." The idea of the brain—of having one, of there even being such a thing—is, if anything, artful.

What a feeling—to do so much with so little, to perform the immense complexity of human experience on the stage of the organ disarticulated. What a way to be human. Art must be operative here, inherently.

As "the mediator of the unspeakable"[3] and having the appearance of something "being rounded,"[4] art in the first quotation reveals the hidden, and in the second aestheticizes and satisfies. Accordingly, once taking a form recognizable as the brain today, an artwork can parade how our personality is a set of neuro-chemical compositions and then crack a joke about getting some satisfaction that we have just represented our most elegant and most true self. We can be tempted to see the artist as a kind of Leonardo da Vinci brain of genius in making a big, round brain that speaks "the unspeakable," and we can celebrate the human-as-a-brain with a haughty proclamation that its complex (but gorgeous, darling) computational aesthetic reveals its superiority among species. But we could also see a Jeff Koons brain. A Balloon Dog brain. And then we might wonder: is this brain a Capitalist creation? And will the brain as art ever sell to a Russian oligarch for 52 million?[5] If it does, will it, like Balloon Dog, highlight the child's play of the artist (and scientist) and the monumentality of an industrial technique that ultimately reflects back the viewer's face in a polished consumer sheen?[6]

In this book, I ruminate on the approaches and materials used to build creative depictions of the brain—what I call brain art. The stitch, the fabric, the dance, and the splash of paint stage the contemporary neurosciences often in an unexpected way. The artworks can be tender, shocking, ironic, sarcastic, or salacious. Many manipulate the images or tools of the neurosciences to generate their interventions, to forge connections to concrete practices, and to recall how neuroscience is a production and a set of dependencies. Philosophical trends and social trajectories are presenced in unusual animations of material objects crafted to appear as identifications of brains. Brain artworks perform the social shape of the neurosciences even as they interrogate the making of that shape. The resulting novel neural visions beg a multitude of interpretations. For me, they show how we imagine and instantiate the brain today.

To the point: brain art—as a brain re-visioned and re-presented—helps us to re-see the brain scan. In almost every case and by sheer nature of existing, brain artworks interrogate a major function of the brain scan: penetration operating under the guise of a pursuit of deeper truths motivating the subject. Brain scan images, like celebrated artworks, aim to expose what no person could ever know without technical intercession, and they appear salacious when they speak what no person would dare to admit (hidden tendencies, maniacal dreams, secret longings). Brain art and brain scans pursue the shedding of the trivial, the ascension of vision, the consciousness raising of the previously unrecognized/able.

I offer an existential book, romping among the rhetorical, philosophical, and sociological. It's a book intended for a broad readership with some sugar for the interdisciplinary sweet tooth. I am motivated by thinking "the brain" through problems of the body; I give attention to our human ignorance—about our state of living, our pains and dilemmas—taking them as entwined with our technological environments. But I see the awareness of ignorance functioning also as an underlying motivator, not only for the creation of brain art but for any investment in what neuroscience might one day do for us. We crave to leap over what we feel is a vacuous gap. A gap of knowing. A gap of feeling. A gap of capacity and capability. This is a gap no machine has yet been able to fill.

To complement the approach, I offer a phenomenological kind of scan with this book, a magnification of my own responses to a neuro-society probed through brain art. Scattered amid the pages are little "scans" of my own brain, i.e. my own secret revelations on the character of the brain today, its assigned roles, its sensuous meanings. These "scans" light up underlying themes swimming within each chapter's analysis of particular works of brain art. They illuminate the frailties of bodies, the instabilities of ontologies, the confusions of technological complexities, and the hopeful assurances of neuroscientific investigations expressed as art. I have scattered these scans strategically to interrupt the reading. They should be like blockages in the flow of things. Little blood clots. And I hope that they accentuate a bodily lens for seeing the brain.

Between the "scans," an argument is presented: brain art simultaneously applauds and laments just how much the brain has been abstracted and transformed into what I call "The Brain," which is a grand popular and fictional brain. The Brain is a figuration of an existential savior in neurobiological robes, an abstraction and an ideal manifesting across all kinds of technical worlds geared toward (mental health, biomedical, professional, personal) improvement and riding the successes of the neurosciences in the past thirty years. We now confront, I argue, The Brain of Power and Might, a social expression aiming to expand knowledges and remake our lived realities. Thus, the abstracted The Brain is not merely a rhetorical operation in the colloquial sense of "rhetorical," meaning The Brain, as I conceptualize it, is much more than an appeal justifying our renewed interest in Sudoku. The Brain is "rhetorical" in the larger, motivating sense: a material and discursive construction re-ordering how we think of ourselves, guiding how we make economic decisions, altering our educational programs. The Brain is a techno-materialist form of subjectivity, leading how we decipher and, indeed, The Brain is the very thing that impels us to decipher which child in the day-care center is a psychopath and which is "normal."

The Brain's magnanimous, solitary "I Alone-ness"—which is a kind of cultural cresting of a philosophical wave started by René Descartes' declaration of "I am" after his audacious transposition of the biblical God's name ("I Am") onto his own tender human mind—concentrates agentive power today in the organ called the brain. The force of The Brain, like a global tsunami, can be detected at airports when security officers are trained to smile "because of mirror neurons,"[7] at local gyms when posters about "endorphins and serotonin" motivate the cliental,[8] and at office complexes when neuroscience studies about "stress triggers" and comfort perceptions are sold to managers as able to "optimize the office" and improve core efficiencies.[9] If artworks incorporating brain imagery are any indicator, then The Brain's contemporary dominance is so extensive as to be understood immediately by the museum-goer as a witty and insightful choice. The brain, only once indistinguishable from The Brain, proves compelling for the museum. And the social exposé through biological frontage is quite a bit of fun for the artist to showcase and the critic to expound.

The interpretive multiplicity of The Brain of the museum

The Brain extends into the arts, certainly, but the sociality embedded in the biology does not escape critical commentary once presented as art. A brain as art interrogates the broader abstraction, i.e. The Brain. The little organ in the museum necessarily shifts attention to the importance of neurobiological investigations but also compels thought about the unnoticed norms and routines that contribute to its particular formation and shape which human experiences are deemed valuable for further investigation. Deciding on the meaning of a brain scan image, we realize when trying to do it on the spot, depends not only on how it was made but on whose brains were compiled to make the image; equally, the meaning depends on whether we believe that we see a cause or an effect there, as well as on things like how we imagine the contours of well-being and how we choose to then apply those contours to the bright red blobs seducing our aesthetic sensibilities.

It is hard, for me at least, to enter a brain art exhibit and resist the easy correlations. I want to see people as "brain problems," that is, to see those brain artworks as brain organs. I do not always immediately see those artworks as compilations of processes, materials, and social presumptions. But when I manage to slip away from the simple one-to-one, I see more. I might even get a little snide about the immense role of the social in the gallery space. I wonder, for instance, why we feel the need to predictably make chit-chat in art museums about our deepest core while holding canapés. Is it a biologically rooted compulsion, like we are somehow "wired" for canapé conformity? Then I remember that the metaphor of being "wired" is itself wrapped up in a social sphere bathing in the delight of new media environments bubbling with computational discourses,[10] just as canapés are baked of an 18th-century culture of elitism where feeling the need to display refinement at cocktail parties popped up through tiny, aesthetically pleasing bites.[11] In fact, the *amuse-bouche* could well be applied to the snitty, witty remarks made in the museum about a brain artwork's philosophical entrée. In some sense, we are always ingesting sociality, even when we pretend otherwise. However, if we pretend to see only the social, then the brain on the wall seems to sneer back at us; the brain speaks like a French film director: *did you imagine that oxygen did not flow through the frontal cortex? That the imaging machine did not capture anything but the scientist's negotiations? Pff. Be real. It is a French brain! Of course it is in touch with the world! By the way, they're serving an excellent 2010 Bordeaux in the lobby.*

Brain art swirls together the nature–nurture binary, not allowing the triumph of either over neuroscience imagery. Take, as an example, one iteration of brain art: a brain sitting on a pedestal in the museum. Do you see it? Encountering such an object compels viewers to consider the uniqueness of the individual while suggesting universality to the human experience and, of course, a common biological composition. The work wants the viewer to imagine the thoughts of this organ and, thus, offers an inherent recognition of our diversity and difference. But the person there, in any specificity, is obscure. The description of the artwork, we can imagine, reads simply: "study of a brain." Whose brain is it? Nobody can tell. Not at least until the brain is united with a body, takes a deep breath, and that body starts to speak. So the "nurture" part may not be immediately present, yet it snaps to attention when the viewer notes the celebration of the organ in this institutional setting and the cultural dynamics at play in deeming it worthy to be called art. Nurture again surfaces in the striking absences, i.e. lack of engagement with an exterior, such a body's accent, gesture, clothing, etc. In this way, a brain in a museum points toward the body as well as how we ignore the body.

A brain on a pedestal necessarily directs attention outside of itself. It asks the viewer to look around. What we have more basically then is a fast line in one direction followed by a quick reversal: the object is elevated—thinking as a divine form brought into appearance and presented as manageable in neurobiology—put there to demonstrate the powers of a sociotechnical society and suddenly making a 180-degree turn, making known all of the reductionisms. The artist is reduced, we assume, to a brain as much as the audience, and the museum appears interested in controlling thought. Here another reversal lingers within the brain's solitary confinement: art today, we might be tempted to think, needs only to find "a brain" and requires no calloused hands or sore back or squinting of the eyes to stage material revelations. So this artwork offers an ironic gesture: the artist tempts us, cheekily perhaps, to forget about making art; assign all of human understanding to the scientific inspection of brains; ignore the artist's love for moving the body, for feeling thrilled, and for discovering new insights by twisting the hands. This reading would not imply so much that the brain is somehow standing in for the artist—probably not being proffered with a yawn as yet another claim about the artist being a genius brain—but could rather be understood as a sarcastic gesture. This brain on a pedestal stands in for the institutional requirements for art and how the artist must submit the mind to the brain at this present time and reduce the body to a brain.

Of course, a brain presented as art could just as well be interpreted as a transhumanist encouragement to step finally beyond the body, make viable the assumed independence of the organ in neuroscience discourses. But if we ever feel excited by that, we would be feeling our own flushed face. That's when the brain in the museum compels consideration of how exactly human consciousness got to a point where it could turn back to its own material form/ation as an object of self-reflection; what good is this reflective ability anyhow? What is the evolutionary benefit? Here, the brain, as art object, points us toward origins and reminds that we are limited by the capacities of the organ, even if it is being held there on high, literally on a pedestal, object of visual fascination. If artists ever show us the future, then there it is. What can be done with this little fleshy thing?

This is not to miss that a brain in a gallery must necessarily foreground the role of technological mediation in human vision and imagination. Shaping art, we remember, is material intervention. We see something new through shaving down. We are technological. Circling back to the brain sitting so nicely and quietly there, we might then celebrate the neuroscience-art project as another expedition into the aesthetic, just as much as into the unknown. Or we might laugh at the chilling silence of the exhibit and maybe feel a little perturbed: *That's it? … All we can do is stare at the thing? … What more do we learn here?*

In visual arts, the brain positioned as the art object cannot but make the knowledge mechanisms of the sciences into a political subject. Creative approaches to the brain not only have the capacity to undercut the popular celebrity of neuroscientists as the providers of amazingly rich insights into the human through specular intervention and expert intersession, but such artworks can also function, at times, to undercut any scientific elitism positioning artists as mere performers who are not able to access "the real" of the human like scientists. Artists, in re-presenting the brain, take back some of the right to material thinking. They turn—like a wooden staff becoming the gorgeous leg of a Massoud chair—all technical questions around and around into something sensuous, which asks us to wonder what materials do to us and what they can do, both within us and for us.

The brain in the museum makes a snide remark. It snickers about who gets access to institutional laboratories and institutional knowledges simply by being so dead and anonymous. So obvious. So there. It does this, ironically, while still being an object secured by an elite museum authority that only allows select artists to showcase their work, according to trends, marketability, identity branding, and the like. And yet, the little brain in the glass case on the pedestal in the museum also asks us to question the artist's imagination for the exhibit. Putting a brain in that space is, after all, a recycled one-trick pony—ripping a well-known object out from its context and displaying it like Marcel Duchamp's urinal.[12] In the case of a brain, there would be no need to slap a signature on the side though; but maybe the artist could use a scalpel to scrape in a signature or maybe use an electronic scannable tag to highlight the amazement and the fright of the seen and the unseen of neuro-technological intervention. Who is in control when the brain interfaces with digital media? Even then, the artwork, like Duchamp's urinal, may not be viewed as immediately controversial, not considered "obscene," not deemed unfit to be displayed and called "art."[13] We are too much of the Enlightenment, too adoring of Descartes, too obsessed with datafication, too trusting of algorithms, too willing to self-technologize, or too fast to apply scientific discourses to identities to see a brain in a museum as anything but stupendous.

No single work of brain art that I know of, nor any artwork discussed in this book, actually takes a human brain from a cadaver and places it on a pedestal in a museum, as described above. But the discussion unpacks a little of what making a brain into an object of creative expression can do. Of course, whether any work of brain art proves to be seriously and discernibly reflexive about its power associations or not—as opposed to narrowly invested in the riches of celebration around the neurosciences and The Brain—is a matter of forms and formulations. However, even if a work of brain art does not appear to critique its own social programs and aesthetic dreams, then it still cannot override its main functions: to illuminate how and how often the popularized appeal to the ultimate defining factor for the human—The Brain—shows up. To shine a light on how we love to see it and what we hope for it. And in doing that, it enacts some critical reflexivity.

The social philosophy of The Brain

What is at issue here is The Brain and why we adore it. As works of brain art make evident across the chapters, a strong-headed, larger than life The Brain acting as the most real guiding force does not trouble us very much. Through analyses of brain art, I hope to show that The Brain eases disturbances about the contemporary conditions of our existence. The Brain acts a priest of our personhood. It comforts us about the untenable future. When we wake up at night and worry about our children and stare at the ceiling, The Brain turns our minds toward neurochemical engineering and the promise of increase neural plasticity with "brain exercises." Some path forward. The Brain is pragmatic. Accordingly, The Brain is enshrined in technical tongues and geared to grapple with attention-sucking digital media, overcrowded cities, unrealistic corporate expectations, financial limitations, among other burdens.

The Brain is voiced as a shepherd for our decision-making processes, and it is made apparent in promises of new insights about how to care for our thorny emotional lives. Discovering that I can "Train [My] Brain Healthy,"[14] that I can "Get a Man Addicted to Me,"[15] that I can "actually build the brain structures we use for relating lifelong," that I can increase my "Emotional Intelligence,"[16] or that I can expand my "mental workplace,"[17] or that I can "boost my memory" if I only follow a "brain-healthy lifestyle,"[18] constructs clever, if not exaggerated, associations between technical-scientific knowledges and the life of the body. This magnanimous neuro-spin is a whirlwind; it blows a lot of hot air, even as it propels the neurosciences into a monitoring and management role. And all of these happenings compose an abstraction and a mode of human engagement: The Brain.[19]

The Brain leads the way and may well stage the coming-into-being of another brain, a digitally integrated neuro-technological brain, which would be another "disciplinary technology of the body," as Michel Foucault could surely see today. The Brain encourages a body to self-soothe with a brain; multiple, diffuse sets of actors (technical and corporate, alike) stage the organ as the means to achieve greater flourishing through adherence to regimented habits and routines dictated through appeals to neuroscientific evidences and jargon.[20] With The Brain securely at the helm, the mind seeks for the body a better life, and on behalf of living experience, The Brain promises to plow through the psychological storms. The problem, of course, is that The Brain shovels a wide-ranging technical-scientific sphere, and quite complicated neuroscience, off into the gutter. Or perhaps a better metaphor: The Brain pushes a brain organ through a butcher's meat grinder. The Brain wants to cook up neuro bacon burgers with *Nature Reviews Neuroscience* for the masses on the money grill. It grinds down the neuroscientific kaleidoscopic through a black-and-grey prism-ideology dedicated to ingenuity, increased productivity, and total efficiency in a clean bar chart. Put simply, The Brain is a booming late Capitalist preacher of neuroscience as much as a turbojet for transhumanist aspirations.[21]

Smash Capitalism together with souped-up transhumanism, and we move into a time, perhaps not far off, when we really do confront another "disciplinary technology of the body." The rewarded life or the life to be punished is a neural measurement and an apportioning. For those who require punishment, The Brain, as a kind of mobilizing concept and technologized future, can offer a "combination of more subtle, more subdued sufferings"; likewise, for those who are rewarded or need to be rewarded, The Brain can secure "the elimination of pain" along with visual and social rewards.[22] Both live "under a seal of secrecy"[23] being guided by a technical process that approaches the brain organ as direct intervention for disciplining populations. We may not yet be at this dystopian vision yet; Foucault's "disciplinary technology of the body"[24] may not now be a brain totally vulnerable to bio-technological alterations or subliminal suggestions, but The Brain already obscures the conceptual difference between that day and today.

Believing that I can reprogram my brain with a mindfulness app so that the same organ causing the hyperactivity can be made into the solution exposes a technologizing function and mirrors the technological bind. In other words, using a meditation app "to calm you down on the same device that stresses you out"[25] demonstrates how The Brain is both problem and solution. The Brain rehearses the freedom and prison discourses structuring how we have thought for most of the past century about technology.

The philosopher of technoscience, Jacques Ellul, clarifies the point. For him, a technological bind exists because the means of freedom through technological realization becomes itself another site of technological struggle. This, too, must be overcome technologically, and often through repurposing or redesigning the same technology. The adaptations cause yet more problems, infuse other kinds of limitations, and those then cannot be abandoned for the risk of returning to the old problem or imploding the entire apparatus, which is now (re)networked. Thus, once again, the technology must be extended and re-devised. The action seeks after freedom but, like an absurdist play with the lead character clinging to a rope while trying to climb a mountain, spins around and around in circles until the character is imprisoned by the rope and ends up hanging himself, making a fool of the technical genius. As Ellul says, "The very fact that man can see, measure, and analyze the determinisms that press on him means that he can face them and, by doing so, act as a free man"; however, ironically, to understand the notion that there are "necessities and determinisms on all sides" means that the measurements then determine the course. Consequently, to see "the real nature of the technological phenomena" and how it "robs" of freedom is to confront a world where thinking about more freedom is always already a technological thinking, such that to think one's way out of the machinic for freedom is to be already defined by the machine and operating not only on its terms but inside of its structures of thought.[26]

In this deterministic view, we encounter Heidegger's thought. For Heidegger, the "essence" of technology is a remaking of the world, including of human experience.[27] As philosopher Iain Thomson explains it,

> Heidegger is really expressing the paradox of the measure; height is not high, treeness is not itself a tree; and the essence of technology is nothing technological. To understand the "essence" of technology, Heidegger says, we cannot think of "essence" the way we have been doing since Plato (as what *permanently endures*) … we need, rather, to think essence as a verb.[28]

That which becomes intelligible, in other words, happens through some engagement with technology and changes because of that engagement. Logics take shape and "reality shows up for us" through what can be measured and accounted for by technology.[29] Replacing the brain organ for the word "technology" in the sentence above illuminates how the brain becomes a technology. The Brain absorbs what Heidegger calls "the essence" of technology, becomes subject to the same routines of recursive problem-solving and re-technologization. The Brain infuses intellectual and social life, influencing what The Brain then does, i.e. how The Brain as an abstraction develops and functions in society.

The technological flesh becomes apparent in The Brain. The brain has long been treated as a machine, but once it becomes a PC, its software needs updating. A shift to bio-integrated neuro-technologies would only be further literalizing Heidegger's idea that humans are "enframed" or racked by technological devices and forced into limited ways of thinking and modes of being as a result. The solution, as noted, would be The Brain and new brain technologies sold on its abstracted and conceptual back.

The Brain, in brief, can be understood as the expansion of technological logics—programmable control, efficiency, optimization, managerial oversight—over and onto the body in a direct and scientific way. This expansion—already itself coming out of past technological evolutions—(re)colonizes the body. For Heidegger, a brain organ conformed to such impulses, that is, a brain subject to The Brain may well represent the most cruel effort to refashion, and in fact destroy, an ontological "essence" of human freedom that could otherwise be sympathetic to alternative environments, bring about fresh accountabilities, and open novel modes, namely, be adaptive without technological possession of human destiny.[30] Put in terms familiar to Jacques Elull, The Brain would be a meta-expression wherein the brain organ is made to absorb all technocratic impulses; The Brain embodies a way to finally submit every aspect of the human expression conceivable to technological integrations. The Brain would, of course, be promising total freedom in the process. But the reshaping of the brain into the image proffered by The Brain would nevertheless basically play the bass-boom heartbeat of the technological bind.

With The Brain, the story, if we want to put it that way, remains the same. The brain organ's divine status—positioned as our Creator made tangible—is worshipped but results in some tangible imprisonment. Turning to The Brain for answers and bowing down reenacts the life of the Puritans who "professed to love liberty" but found themselves "restrained by strict laws" in an effort to garner favors from God.[31] In this case, though, the worship would be a self-worship that is also a worship of the social and political configurations embedded in neuro-technological forms. And here the story has an opportunity to diverge: The Brain, unlike other divine images historically offering human guidance, is not a permanence nor independent of human action. The Brain transforms along chains of encounters with the brain and, we hope, revises through a multitude of material expressions, including the artworks in this book.

Digesting creative expressions of the brain organ and setting them in relation to today's technological society helps us to step back and see The Brain—to reveal its scaffolding and make apparent its implications. In the process, what should not go unnoticed is how brain artworks hold transformative potential. There, they help us to understand just how often The Brain speaks into the deepest realms of the human organism, whispering the internalization of agency, countering existential deflations, promising a new life; but should we listen? How new can we become? What do we want for ourselves anyhow?

When we feel alone and demand an answer—a reason for our life, suffering, and death, and try to find a way to come to grips with our situation—The Brain extends a hand and declares itself a steady route toward comprehensive composure. But it is simultaneously optimistic and damning. It is a frontier that inspires but can only turn to its own conceptions for better ideas about where to venture next. It is a model for overcoming the limits of human capacities shaped from those limits. Clues to this arise in the artworks that show The Brain cooperating with affective circulations in Capitalist economies. The Brain, like a new Corvette, hums over the drone of existence in suburbia. It doesn't matter if it drains the wallet, because on an emotional level, it feels pretty cool.

If The Brain is going to help us swim through life, then brain art helps us look down to see the moves that we use to stay afloat in the deep water.[32]

My approach to brain art

I understand brain art as a technê, or "productive technical art,"[33] that rivals the technê of the neurosciences. Even as artworks manufacture their own allegiances to The Brain and bow before the epistemological powers of the neurosciences, they also compose yet another brain. The fluid, messy, sticky, bright brain of brain art interrogates and revels in—glitter bombs—the psychological, social, and political functions and dependencies of the popularized The Brain. Focusing on brain art, then, allows for a different approach to "neuro everything." Brain art presents a different means of access; material revisionings offer sensual, embodied appraisals of the technical assurances pasting over our ontological vagaries. Brain art tests and interrogates the on-going drive to add clarity and definition to life through neuro-technological means. But if brain artworks entrench any resistances to neuro-dominated domains in creative ambiguities and multiplicities, then they also take seriously the idea that brains can be put forward as tangible solutions to a body grappling with strangeness, deterioration, and uncertain futures.

The artworks discussed in this book were chosen both because of their prominence and their intrigue (from my perspective as someone interested in the shape of the neuro-humanities).[34] Many of the artworks appeared in well-known museums or have won awards for experimenting with neuroscientific tools and laboratory results. The artworks that kept cropping up, namely in searches of high profile art journals like MIT's *Leonardo*, caught my attention but also enabled me to read the artists' statements and consult the artist's own goals and perspectives.[35] That gets at the question of my approach.

In analysing brain artworks, I follow the close reading tradition common to fields such as media aesthetics and visual rhetoric.[36] But I focus not only on internal aesthetic compositions but also on the material choices and the way that arrangements connect to specific contexts.[37] I ask what are the elements, what is being inter-related, how do the works hold together, how do the materials guide our feelings about it, what are the effects on the communities implicated, and what are the broader philosophical commitments? That is, I work inside the composition and then move out to contexts. I expand my thinking until the work drags me back into itself to revise the reading or suggest another direction.

I actively try to offer several possible readings for each artwork. I hope to introduce complexity. I hope to show how brain artworks are traitors to themselves. Indeed, each can propose an opposing point of view. What emerges is a series of contradictory images: the beauty of neurological networks side-by-side with the gore of cutting open skulls to poke around and see what happens. Ominous oscillations between the strangely serene and the outlandish. Neon pop inside of a formaldehyde filled jar. Sometimes brain art entertains new possibilities for body-world configurations and dismantles totalitarian thinking; other times, it seems maddeningly unaware of how its configurations advertise reductive explanations of the human (you *are* your neurons, dude). Often it does both simultaneously.

What I hope will be evident is how brain art plays along the sharp blade of a double-edged sword with a big stubby finger: down the first edge, it applauds neuroscience, while running up the other, doubts the power of its ensembles. We cut ourselves and bleed. Brain art caresses neuroscience's emotional resonances yet lays bare its social contingencies.

But the reader may encounter at least one sticking point. Some audiences might wrongly come to believe that my philosophical and social critiques mean that I see the neurosciences as inept or as corrupting influences. Yes, I am critical at times; I am not trying to be overly generous in exploring social investments and implications. But I aim to critique what people do with the brain today. I explore the meanings that have been *applied to* the brain and its disciplined study.

I understand that neuroscientists are often disconnected from the strange deployments of their collective work. The neuroscientists that I know desire to understand the organ and often focus on one particular aspect or function. In so doing, they confront the tough task of relating their work to a broader set of circumstances to show its relevance; they have to deal with multiple stakeholders and mediators along the way. So neuroscientists cannot be held easily responsible for all of the ridiculous "neuro-hype," as van de Werff calls it.[38] But that does not mean that neuroscience, as a practice, sits free from philosophical, social, or political critique. Investigating how neuroscience studies position the brain while trying to better understand why neuroscience fascinates us in the first place responds to a culture now bombarded by numerous popular brain books asking us to sing to our babies, eat more blueberries, get out of bed earlier, practice looking people in the eye, play memory games, all because—why again?[39]

I take for granted the idea that neuroscience is popular and cool. People love it. I love it too. I have spent my career thus far writing about the communication of neuroscience and the conceptual frameworks used to interpret the brain. I am truly fascinated. But as a communication scholar, I am also interested in the material and symbolic realms that entwine to inform how we approach the brain. This necessarily involves engagement with the "Critical Neurosciences" and "Neuro-Rhetorics," sub-fields interrogating the many promises and the cultural and political positionings of the neurosciences and its various popular iterations.[40] If ever coming across as uncongenial, this book, be reminded, seeks not to attack the contemporary neurosciences but to compel a serious social and existential appraisal that can unwind implications and better grasp why we salivate over the brain.

My goal here is to locate human tendencies, psychological compulsions, and socio-cultural repetitions—and to do so primarily by looking at artworks that take the brain as object of creative representation. My goal is not to write an art history book, nor do I aim to praise the neuro-aesthetics movement. I feel no need to rehash what art reveals about the ways that our brains work. I do not look, for instance, at what a functional magnetic resonance imaging might say about perceptions of Beethoven's music. This book is not part of the "neuroscience of art" movement. There are plenty of those books out there, and a cognitive neuroscientist would be better situated to detail the anatomical and functional results. I want, instead, to consider how we feel about the brain, how we talk about it, how we celebrate it, how we show it to each other, and then ask if human life with a brain can or should be thought differently.

Putting my own ideas and feelings in conversation with continental philosophy and critical-cultural and rhetorical theory, I ask: what are we trying *to do* with a brain? How does a neuro-everything relate to our festering uncertainty about the state of things, i.e. how does the brain help us to push past existential anguish at not fully realizing what our bodies are doing or where we are going or the effects of expanding technological complexity amid the threat of a dying planet? Will the brain eventually leave us stranded?

That's what this book is about. The chapters touch such themes softly, like fresh wounds, like boils seething on our hands and feet. We need ointment. We need aspirin. We get stomach ulcers. We have heart attacks.

Overview of chapters

The first chapter stresses a time before there was a concept of the brain. Narrating ancient histories of the body from Italy, Egypt, China, among other places, foregrounds the long human search for a way to describe our relations and gives us a point from which to rethink a neuro-obsession. Before the brain concept was as clear-cut and elevated as today, we can see how others figured the body and consider whether we can reverse course to learn from and mirror some of the ways that they composed life. The next two chapters work through six substantial, contemporary works of brain art, i.e. three in each.

Chapter 2 argues that one outcome or expression of brain art is the sensualisation of neuroscience. As a tendency, or a thematic, seen in the works analysed within the chapter, I call this function Neurosensuality. It stresses the tenderness that happens when encountering a work of brain art and within the process of experiencing one's self as a brain. The artworks analysed in this chapter demonstrate the intimacy of technicity. They show how a science of the brain squeezes the heart. They entreat the neurosciences to more strategically care for how neuronal descriptions and interpretations embed in people's lives. They perform how brains get enshrined in homes.

Chapter 3 argues that other brain artworks harbor a tendency to celebrate a technical effort to objectify and universalize our feelings, specifically our strange and sudden or uncontrollable feelings, i.e. affective experiences. These works, in one way or another, invest in direct correlationism and technological transparency. They sparkle with the dreams and wonders of The Brain. Some of the artworks respond to this tendency almost as a presumption while others enact it like serious TV drama. That is, because artworks can be multiplicitous, they have varying dedications. Some actively question if neuroscience can, in fact, detail every feeling and shape a commonality for humanity. Others seem to assert an unqualified answer. Overall, I call the thematic expression noted in these artworks Affective Realism.

Chapters 2 and 3 are offered as case studies. They are not meant to be the final word on brain art nor meant to be all-encompassing in terms of covering the totality of the terrain of brain art. The chapters are, instead, an entrance into a conversation about this phenomena of the brain being made into an object for creative revisioning. The chapters simply set out to formulate an answer to a question about what happens when artists foreground neurobiology and highlight social realities by engaging imagery from the brain sciences.

Chapter 4 then steps back to think what more brain art might do, how it might further develop critical faculties. I spend some time in Chapter 4 wondering whether brain art can more daringly expand ontological imaginaries—offer alternate visions of the human—or whether it leaves us stuck in the mud of objectifying scientism. I specifically aim to point out what has not been clearly articulated thus far in works of brain art: the wildly offensive, the spitting "fuck you style," and the disruptive performativity of the instantaneousness and the generativity of Things staged in a charged political and increasingly technical landscape.[41] In the final chapter, I then pose a challenge, which is intended to be philosophical as much as pragmatic: can we dethrone The Brain as guiding force? What about imagining life without a brain concept?

Only the reader can decide if we are too attached to stability or too invested in technicity or too reverential or too sentimental to re-see neuroscience as a form of creative arrangement, like art, as a construction built of sensual seductions, laboratory imperatives, and sleepless nights. Only the reader can decide whether a neuroscience finding can be an arrangement of 10,000 material co-dependencies negotiated amongst colleagues over cake and coffee. And if we are not so interested in embracing a time, now so very long ago, when there was no such thing as a brain, then how do we move forward with what we have in The Brain to create a more socially sensitive and politically vibrant brain?

Artists might choose to do what The Brain asks them to do: give neurobiology the highest priority, set it on a throne gilded in gold, and answer the clarion call to shrink down all of our human experiences to the crisp technicity of the scan. Perhaps within the glittery glow of artistic exaggeration, under the bright lights of Madison Square Garden—and we shall see—The Brain will smoke out our expectations, despairs, and supplications. The Brain might even one day suffocate its own concept and give life back to the body—and not one body: many little bodies coming together, anew all the time. Then, a brain will look nothing at all like The Brain. But the brain will also not even look like the brain organ. What we will have then—all things in transformative action—will tingle with everything, all things becoming brainy.

Notes and references

1 BOLD is an acronym standing for blood-oxygen-level dependent contrast. See: Seiji Ogawa and Tso-Ming Lee, "Magnetic resonance imaging of blood vessels at high fields: in vivo and in vitro measurements and image simulation," *Magnetic Resonance in Medicine* 16 (1990): 9–18; Nikoa K. Logothetis, "The underpinnings of the BOLD functional magnetic resonance imaging signal," *Journal of Neuroscience* 23, no. 10 (2003): 3963–3971.

2 See "Brain," *English Oxford Living Dictionary*. Italics added. Available at: https://en.oxforddictionaries.com/definition/brain.

3 This quotation has been recycled so often that it is difficult to find its origination; it is attributed to Johann Wolfgang von Goethe, however, I have not found it in his works as of yet. Even so, we will give him credit and trust that he did say this.

4 Quotation by Maurice Denis, "Definition of neotradition," in *Theories of Modern Art: A Source Book for Artists and Critics*, translated and edited by Herschel Chipp (Berkeley: University of California Press, 1996), xv.

5 For an article on Jeff Koon's Balloon Dog artwork, see: Lorenzo Pereira, "Most expensive Jeff Koons Balloon Dog pieces," *WideWalls*, November 1, 2015.

6 The *Guardian* reports that Jeff Koons called Balloon Dog "materialism and monumentality," a quote that inspired the idea here. See Laura Cummings, "Jeff Koons at the Ashmolean review: a master of deflection," *Guardian*, February 10, 2019.

7 This comment arises from a narrative of airport security reporting this to a neuroscientist interviewed by the author in 2013; however, a similar comment is made by Kevin Seybold in *Explorations in Neuroscience, Psychology and Religion* (New York: Routledge, 2016) in the section titled "Neuroethics."

8 Lorne Opler, "Your brain on exercise: the neuroscience behind a good workout," *American Council on Exercise*, October 10, 2018.

9 John Medina and Ryan Mullenix, "How neuroscience is optimizing the office," *Wall Street Journal*, May 1, 2018.

10 Jordynn Jack, *Autism and Gender: From Refrigerator Mothers to Computer Geeks* (Urbana: University of Illinois Press, 2014), 191.

11 See Tory Avey, "Speakeasies, sofas, and the history of finger foods," *The History Kitchen*, February 1, 2013.

12 See: "Marcel Duchamp, Fountain, 1917, replica 1964," *Tate*. Available at: www.tate.org.uk/art/artworks/duchamp-fountain-t07573.

13 Ibid.

14 See the article about dementia titled, "Look after yourself: train your brain to be healthy and prevent dementia," *Careforkids.co.au*, May 20, 2015.

15 Jessica Taylor, "How to get a man addicted to you," LiveStrong.com. Available at: www.livestrong.com/article/193334-how-to-get-a-man-addicted-to-you/ (accessed May 10, 2019).

16 Linda Graham, "Neuroscience of attachment," *Linda Graham, MFT*, September 5, 2018.

17 Eric Tra, "Science of imagination," *KillerInfographics*, October 12, 2016.

18 Elle Kaplan, "4 proven strategies for improving your memory," *Thrive Global*, April 18, 2019.

19 Even in using the term "processing," I myself reveal my own adherence to The Brain, despite my critique of it. The Brain embeds; it is an open secret.

20 See Michel Foucault, "The body of the condemned," in *Discipline and Punish: The Birth of the Prison*, translated by Alan Sheridan (New York: Pantheon Books, 1977), 3–32.

21 Something similar is noted by Davi Johnson Thornton in the opening pages of her book, *Brain Culture: Neuroscience and Popular Media* (Piscataway: Rutgers University Press, 2011), 1–6.

22 Foucault, "The body," 8.

23 Ibid., 10.

24 Ibid.

25 Eleanor Cummings, "Meditation apps want to calm you down on the same device that stresses you out," *Popular Science*, May 10, 2019.

26 Jacques Ellul, *The Technological Society* (New York: Vintage Books, 1964), xxxiii; for discussion of the "real nature of technological phenomena," see pp. 8–19.

27 Martin Heidegger, *The Question Concerning Technology,* translated by W. Lovitt (New York: Harper & Row, 1977), 33.

28 Iain Thomson, "From the question concerning technology to the quest for a democratic technology: Heidegger, Marcuse, Feenberg," *Inquiry* 43, no. 2 (2000): 207.

29 Ibid.

30 Ibid., 209.

31 "Puritan laws and customs," *History of American Women*, womenhistoryblog.com, October, 2007.

32 For discussion on the ways that values drive technological development, see Andrew Feenberg, *Transforming Technology: A Critical Theory Revisited* (Oxford: Oxford University Press, 2002).

33 Chad Wickman, "Rhetoric, technê, and the art of scientific inquiry," *Rhetoric Review*, 31, no. 1 (2012): 21–40.

34 The analyses of works is from 2017–2018; anything appearing after that date must wait for future books.

35 Artists' statements for all works were consulted. In the case of Marjorie Taylor (Chapter 2), Jane Prophet (Chapter 3), and Laura Jade (Chapter 3), I was able to have personal or email conversations with the artists. In all cases, however, the relevant journals or artist websites that were consulted are cited.

36 For discussion of the processes and aims of media aesthetics, see Roberto Simanowski, *Digital Art and Meaning: Reading Kinetic Poetry, Text Machines, Mapping Art, and Interactive Installations* (Minneapolis: University of Minnesota Press, 2010).

37 For discussion of visual and material rhetorics, see the following three examples: Greg Dickinson, "Joe's rhetoric: finding authenticity at Starbucks," *Rhetoric Society Quarterly* 32, no. 4 (2002): 5–27; Cara A. Finnegan, "Recognizing Lincoln: image vernaculars in nineteenth-century visual culture," *Rhetoric & Public Affairs* 8, no. 1 (2005): 31–58; Peter Simonson, "Rhetoric, culture, things," *Quarterly Journal of Speech* 100, no. 1 (2014): 105–125.

38 Ties van der Werff, Jenny Slatman, and Tsjalling Swierstra, "Can we remedy neurohype, and should we? Using neurohype for ethical deliberation," *AJOB Neuroscience* 7, no. 2 (2016): 97–99.

39 For examples of neuroscience sold as transformative and geared for popular culture, see Daniel G. Amen, *The Amen Solution: The Secret to being Thinner, Smarter, Happier* (New York: Random House, 2011); Sharon Begley, *Train Your Mind: Change Your Brain* (New York: Ballantine Books, 2008).

40 See Suparna Choudhury, Saskia K. Nagel, and Jan Slaby, "Critical neuroscience: linking neuroscience and society through critical practice," *BioSocieties* 4 (2009): 61–77; David Gruber, Jordynn Jack, Lisa Keranen, John M. McKenzie, and Matthew B. Morris, "Rhetoric and the neurosciences: exploration and engagement," *POROI* 7, no. 1 (2011): 1–7; Jordynn Jack, "What are neurorhetorics?" *Rhetoric Society Quarterly* 40, no. 5 (2010): 405–410.

41 The lovely saying "Fuck you style" was seen spray-painted on a wall outside of Copenhagen's airport, and the author, thus, considers it a diffuse popular statement of rebellion; although it may also be a street name, it is can only be found in *BraskArtBlog*, which documents street art. See: www.braskart.com/tag/fuck-you-style/.

1

ANCIENT HISTORIES OF THE BRAIN CONCEPT

Tucked between two bulbous hemispheres of tissue, deep inside the head, rests a tiny, hard pinecone. The pineal gland. Fascinated by its strange twirl, Descartes declared the gland the "principal seat of the soul, and the place in which all our thoughts are formed."[1]

Pinecones, like pineal glands, are a magical shape. They are an inward-turning spiral. They direct all to the center and up to a single point. Their aesthetic charm and simplicity seduce. They reiterate a social semiotic tied to the environment and the body: they twirl toward the sky to a sun that burns in a hot circle. We are spellbound by the heavens. The stars circle us at night. The universe wraps around us in the ghostly span of the Milky Way. We gape in awe. The aesthetic from inside of the earth is a giant, magnificent spiral. Then, with luck, we discover the pineal gland, just there, absorbing the symbolic weight of our material world. It is a fitting representation of that feeling that we have—a vibrant connection to a geometric outside. And its location seems perfect for endowing us with heavenly significance: center of the head, behind the eyes, like a topper on a Christmas tree. Bodily flesh in the gland mirrors the rotunda of the heavens.

Today, contemporary neuroscientists believe that the pineal gland secretes melatonin, regulating when we are awake or asleep.[2] This is a fortuitous connection to ancient imagery ascribing supernatural awareness and spiritual awakening to the pinecone. Packed with seeds, the cone grows into a majestic tree and can, in fact, generate a forest covering thousands of acres. Perhaps this connection to fertility is why the pinecone rests at the heart of the Vatican's courtyard, at the top of the Pantheon, inside Roman governing buildings, in the hands of Sumerian gods, and on the staffs of the ancient Egyptians.

Whether the pinecone is anything other than a symbol of aesthetic perfection and a mathematical elegance, whether it exposes an underlying mystery about earthly materiality or holds many dark secrets of the mind, teeters on the edge of

FIGURE 1 "Court of the pinecone, the Vatican," Copyright David R. Gruber.

a question about what the body is, how we imagine our place in the cosmos, and how we collectively represent our Being. But whatever we think of the pinecone, we are desperate to find some greater meaning, to see and feel something bigger … but what?

Scan 1: seeking a One

Efforts to determine a founding principle, to unearth a One that explains human experience, are witnessed across history; they are not only repetitive exercises but central components of affective struggle in religion, mysticism, and the arts—and now in neuroscience.

On the left side of the body, we reach out with one hand to touch the sky; we roll the opposite hand to the chest, turning inward to reflect on personal experience. Like a statue in dramatic pose, we perform the dual pursuit of human understanding—outer sensation and inner reflection. We enact a little version of *The School of Athens* fresco where Plato points to the heavens and Aristotle to the earth. We mull space and time while taking apart the washing machine.

In *The School of Athens*, the temple of philosophers is a luxurious marble hall signifying a higher purpose—the search for truth and beauty. For us, neuroscience laboratories can act as another kind of temple, signifying a technical search for universal mechanisms explaining our behaviors. If the perfect geometric architecture of *The School of Athens* suggests an "onto-theological order" to the universe,[3] then the mapping of the brain constructs another philosophical and material architecture, equally fascinated by harmony, grandeur, and governance.

The paleolithic brain

Once upon a time, the brain was nothing.

Paleolithic people who marked out the hunt in mineral pigments on cave walls chose to represent majestic bulls and horses, portraits of strength and beauty. They painted the necessity of killing food, the wildness of living ecologies, the intense struggle of life, and the splendor of ceremony. Nowhere do we find a brain.

Inscriptions of the human body in paleolithic cave drawings flash out of the darkness, but they glow from pigments the same as those used for birds and fish. The shapes seem rarely much different—a line extended out to an arm, extended further to an arrow that shoots off into the sky. The entwined aesthetic cannot be generalized across absolutely all cases, but singularity or division for the human body appears out of the norm.

The striking human figures in the Gwion Gwion rock art, previously called the Bradshaw rock paintings of Western Australia, offer one interesting example to consider. They depict early human life but only inside of a permeable world. Elaborate dress and family groups swirl amid snakes and kangaroos, the sun and the desert. This record of coexistence with animals within voluptuous environments survives, and we can see in those images a type of thinking where life-worlds thrive as entwined expressions. Stress on the individual perspective distinct from any community, as scholars have noted, is not readily present.[4]

Of course, what it might mean to assert an individual perspective in an art image remains open to debate. That is to say, it is not entirely clear that the so-called "individual" would be the same kind of entity across spans of time and different cultures. Locating the "individual" would require criteria and a line of argument. Generally, we can assume at least a few universalizable aspects of "the Self" as a formation. To have a so-called "Self," one likely needs awareness of a basic difference (me vs you), awareness of change or ideational and bodily development, the time and capacity for inner reflection, and some socio-cultural orders or oppressions and repressions wherein the "I" can be self-located in a hierarchy or realm of Things. But historical processes, technological change, mental adaptions to social conditions amid narratives that explain what the group *is* and what the hell an "I" is supposed to be doing in it, probably have wide-ranging effects.[5] The Self is a concept breaking away from unification, dedicated inherently to diversification. Yet, in many ancient rock art depictions, we get the sense that breaking away is not the core idea. The human seems quite at home amongst a multiplicity of living things.

The idea of the individual in the paleolithic era remains elusive and mysterious, as does how the body relates to a concept of personhood. But perhaps we can look again at Gwion Gwion for some clues. One image is especially intriguing, a little haunting. We see there a burst of brilliance emanating from the head, which appears to show thinking; but the design could also take the shape of a headdress, or maybe it shows the sun, or maybe it references the cosmos. If feathered or sun-struck, the human body remains artistically enmeshed in environmental Being. This is not to say that the human face is not special or visually select in the image, but rather this is to say that thousands of years ago its chalky glow may not have been seen as self-generated mental power nor anything of the body as meat.

Chalky lines fatten and blur, making a semi-circle that spans the head of a ghostly figure. A halo complements the headdress, reminding the viewer of a tie to the earth and sky. The impression, to today's art critic, probably automatically reminds of (much) later Renaissance paintings of saints illuminated with God's divine glory. Here, the "halo" can be conceptualized as "a fossil of human experience" in as much as it is an aesthetic expression of a social designation for the person pictured;[6] but anything like this remains speculation and anything more than this remains shadowy.

FIGURE 2 Gwion Gwion rock art, Australia. © Getty Images

Of course, the halo, a distinct and important symbol in various Western artworks, symbolizes a connection to divine dimensions, displaying the communal belief that human reason equates to human exceptionalism. The suggestion that Western art practices are related to something here, something much further back, a kind of compulsion to seek the divine or to embody higher supernatural realms sensed and storied within communities, raises intrigue. But in the Gwion Gwion image, the lack of definition in the face and the absence of any directive gesture calls this narrowly Eurocentric interpretation into question. To go further: reading this rock art through the lens of the Renaissance and its distinct expressions of divinity constructs another Western colonialism scraping away the meanings attributed by ancient peoples and including now those local indigenous communities in Australia.

That these paintings were ever called Bradshaw rock art—in honor of the 1891 colonialist Joseph Bradshaw who refused to attribute them to aboriginal peoples— says something of the arrogance and power of colonial absorption and the failure to understand art on its own terms within its original contexts. A renaming exercise in Australia, now increasingly presenting these images as "Gwion Gwion rock art" across museums and archaeological sites, offers hope of overcoming the historical circulation of Bradshaw's false notion that these paintings were too sophisticated for aboriginal peoples in Australia. Bradshaw was likely unable to see beyond "a 19th Century linear worldview in which societies progressed from primitive to advanced."[7] Indeed, the paintings clearly display "a developed culture," showing human figures in ceremonial gowns, engaged in dance, surrounded by animals, arm in arm. And their age—at minimum, around 5,000 years old and at most 40,000 years[8]—surely stunned or confounded Bradshaw, who probably understood his own history of rational-minded civilization looking back not one-quarter of that span. Even so, some lingering controversy around who painted the images remains in Australian society, despite academic papers putting competing theories to rest.[9] In short, no direct or easy comparison to g/God's divine glory is necessitated here, no correlation to the social and political forces of the Enlightenment is needed when we view the faces in Australian rock art.

Because aboriginal rock art is designed for a community, interpreted within it, and made for a defined time and place, it remains difficult to say with any confidence or dimensionality what, exactly, these Gwion Gwion rock depictions once meant.[10] As McNiven notes, when asked about the rock art, the aboriginal people in the Kimberly region of Western Australia tend to say, "'They are before our time," but they sometimes recall a myth of a beautiful bird (Gwion Gwion) that was believed to live in the mountains and paint them with its beak.[11]

We might then see the thick line around the head coupled with the headdress as a symbol of the bird or an expression of it. Or we might return to the alertness of human consciousness and tie that to the activities of the bird, the great height from which a bird can see. At that juncture, we might again encounter the image as a person thinking vast thoughts, staring right back at us with a kind of historical demand to think again about human life, to reassess the value and power ascribed to so-called "civilization," and to realign the cosmos and see all of life moving together—both in geometric and figurative ways.

The image could, of course, be viewed as a representation of the aboriginal awareness of a majesty or magicalness of the world. But regardless, what strikes the viewer immediately is the depth of that blur around the head. The thick line provides some reason to believe that it stands for the figure's own depth of thought or maybe depth of a perspective. The figure looks out at us. Perhaps it is a self-portrait. But what does the person pictured there see? Does it see me, or does it see an obvious "we" amid the creatures flying above and swimming within? Or does it see right through and beyond any individual viewer? Perhaps we, the viewers, can look back with this look, this looking, and not look at it as an image of someone looking at us.

Consider the hyper-simplified shape of the eyes and nose. Small and apotropaic. The human face holds an inescapable resemblance to what appears to be a fish swimming in and around it. Same shape, same beady eyes. And then, inside the body, we find another fish or maybe it is a satchel that looks similar to the animals on the outside of the body, which appear to be kissing. They touch mouths for the love of art or the unimportance of differentiating the lines within a presumption of co-being.

The head pictured in this ancient rock art, if at all offering a brain-like symbol, if at all suggesting human insight, might well denote an alertness alongside mutuality and environmental similarity. If indeed the figure is drawn to display a sense of consciousness, then the figure, as rock art, appears to be thinking out from the affordances of the surface of the rock, as deeply as the pigments allow, and those pigments we can see are made of the local rock whose geologic time is immensely more ancient.

The brain here is nothing. When following the lines and counters of the rock, we are unable to move very far beyond the magnificent impression of a vibrant ecological life tarrying alongside, and indeed reliant upon, the resources of the place.

The brain of ancient Egypt

The ancient Egyptians yanked the brain out of the skull and threw it away. When preserving important organs of the body for the afterlife through mummification, they apparently thought nothing of the poor squishy brain. Anatomical violence—a cortex hooked and splurted out through the nose.

Like many ancient cultures, the heart and blood dominated the collective imagination. The dramatic beating, the bleeding. God help us! The rush of adrenaline. The flow after an outrageous battle. The swell of the chest. The clenching of the chest. The heart makes for good associations, and it makes good on them. Pounding away like a timebomb, there is little question as to why it garners fame. Human mentality capacity, as Aristotle argues several hundred years after the New Kingdom of the ancient Egyptian era, flows absolutely and only out from the heart.[12]

But this historical occurrence does leave a strange question about the brain unanswered. What of head injuries damaging thought processes? How could the brain—the one place that we now examine in excruciating detail to understand ourselves—be so blatantly ignored?

As it turns out, the brain was not ignored. Many ancient people did have interesting ideas about its function. Examining the oldest known medical records of an Egyptian doctor, it is clear that the brain was correlated to movement and to several thinking capacities. An injury to the head was noted as able to cause serious dysfunction.[13] However, the brain as we know it, meaning as the organ completely responsible for thinking and feeling, was not how the brain was known. It is difficult to say that the brain was a thing in any sense the same as today.

The age when *there was no brain* is a time worth revisiting in the age of *neuro-everything*.

To think the brain as playing only a small and mysterious role in speech and motor function, as the Egyptian sage Imhotep described in ancient Egypt (1600 and 3500 BC),[14] should not encourage any joke about ancient people's ignorance; the observation might rather spur us to cultivate a deeper desire to grasp their broader concept of the world. Asking what their ideas offer to us today is more reasonable and ethical than scoffing at their supposed failure to inscribe a brain with mastery or to make a man like Imhotep into an archetype for a genius brain. If they were to look ahead to our present and watch us obsess over the brain as the computational Command and Control Center, then they might well scoff at us.

What we can discover in this ancient time is not only that there is no need in certain social and political conditions for the brain to be considered a source of thought, but that thinking itself, as a bodily production as well as a social imperative to drive innovation, can be differently thought. The social marches on and with glorious effect in ancient Egypt. Pyramids and priesthoods do not need a concept of the brain. Astronomical discoveries and architectural achievements do not need to be creations of brains. Thinking at that time is not something produced in isolation, and Imhotep proves never to be thinking alone, never cast as a generative brain.

Scan 2: the brain is a dream

Skin peeled back, skull sawed open, we look to conjure a theory for ourselves. We see it. The Brain. Exposed. We could be looking at a 16th-century Vesalius drawing of an anatomy lab in action. Or it could be a black and white film with a sexy Hollywood da Vinci at it again, slicing open a cadaver stolen from the morgue. We want to see the mad genius on a journey and pretend that ignorance will be banished forever, once and for all, by the scientist hero. But there is little clue in the goo.

We squint, lean in closer. There it is, an alien Thing, quiet and innocent, perched at the center of a team of surgeons who peer down, stupefied.

They are techno-archaeologists mesmerized by an ancient rune. Indecipherable squiggles of squish.

One ancient Egyptian doctor describes the brain as rippling "like those corrugations which form in molten copper."[15] The comparison is visually apt but also indicative of a scarcity of related references. What is it? How should we name it?

We choose a name according to what is hidden but also what wields power, according to a dream of ruling over ourselves, totally, and never again being pulled haphazardly along by a rash of viruses living in the pond or the trauma of remembering the rushing water of a killer flood: The Command and Control Center, The Processor, The Engine. We name it The Brain.

Imhotep's brain

Imhotep, deified as the Egyptian god of medicine, was also called a god of magic. The ceremonial temples in Memphis were dedicated to both with little distinction. One inscription describes Imhotep as "sculpture and chief of seers."[16] The dual label hints at the intimacy between artistic practice and magic. Bringing materials to life was part and parcel of shaping them, a process of making worlds. The artist, the architect, and the doctor were endowed by the gods as human-creators; the source of inspiration, then, was not a brain nor a mind, per se, but a collective, a unity of gods living with the human.

Consideration of Imhotep's immense achievements helps to explain how material and metaphysical life could be brought together for ancient Egyptians. Being the inventor of stone architecture, Imhotep moves Egyptians from mud and brick work to elaborate structures and monuments that would last for ages.[17] The "majestic girth and regal status" of the columns "betokened the power of the pharaoh and the domination by his united kingdom."[18] Imhotep prepared the way for many millennia of government infrastructure, crafting a new monumental synecdoche for the glorious, powerful body of the king and queen, which itself represented entire realms, divine and earthly.

The weight of Imhotep's architectural allusion to power coupled with his ability to climb toward the sky with impressive certitude likely altered social rituals. As Linzey points out, the creation of the pyramids presents a case of Heideggerian "destining," or "the essential innovative and progressive role of architecture as a mode of cultural formation."[19] Imhotep's invention had impact on mental structures in ancient Egypt, allowing stone to correspond to the solidity of empire as well as the grandeur of Being. Accordingly, heavenly significations, long latent in ancient Egypt, took form in and amongst new stone buildings. "By the Fifth Dynasty, for purposes that again are not at all clear, comprehensive anthologies of hymns, spells, and prayers were being inscribed inside the tombs and cenotaphs of the kings."[20] Perhaps the stone had something to do with it. Making what lasts beyond one generation serves religious ceremony while encouraging a belief that an essential life force continues far beyond death. The seeming eternity of stone likely shaped feelings of social and cultural solidity and helped to protect against institutional ruptures.

The scale of Imhotep's achievements, figuratively and literally, is evidenced in depictions of him. He is seen holding an ankh—a symbolic staff shaped like a cross with a circle on the top—which places him amongst many gods and pharaohs. A careful look at the symbol further illuminates why Imhotep was holding it but also how the Egyptians linked human creativity with the metaphysical world.

At first glance, the ankh appears to be a stick-figure human with an enlarged head. The symbolic entanglement of the human head, ruling guidance, and divine power offers one interpretation of the ankh. The sun as a divine symbol in the sky—setting the course of seasons and shaping life—mirrors the rounded shape of the head. The idea that the sun and the head are related finds support in the fact that kings and queens ruled with the guidance and approval of Ra, the sun god.[21] But again, their brains were likely not the issue. The point, instead, may well have been the image of a royal head, which seemed to increase in symbolic allure by the middle period of ancient Egypt, around the time of King Tutankhamun.

As is now well-known, pharaohs during the 14th century BC had oddly oblong heads. This fact likely became a key signification for their deified status. A popularized CT scan of King Tut's head published in *Smithsonian Magazine* foregrounds, if not over-indulges, this historical relation.[22] But the elongated head was, without question, shared by numerous relatives and, accordingly, added visual weight to any claims of royal lineage and to the notion that royals were appointed or determined by the gods.[23]

Contemporary historians, not surprisingly, look to neuroscience to better understand the appearance of this strangely shaped head. What results from a CT scan study of the medical image is the conclusion that the head shape resulted from natural growth. Professor Braverman of Yale University argues that the king had a genetic disorder causing the massive head. Of course, genetic alterations are more likely to be the case given the close relations needed to preserve a royal lineage. Yet symbolically, the heads were, we might say, a divine gift; carvings that indicate the decorative prominence of the so-called "royal head" attest to the lasting symbolic import of the genetic anomaly.[24]

Whether by unintended genetic occurrence or by strategic social choices over generations of pairing a king to a queen, the head, as a symbolic object, forged a strong connection to rulership. The large heads of Egyptian royalty likely helped to evidence the divine on earth. If anything about the body of the pharaoh and his family signified heavenly power, then it was the head—foresight and leadership invested in human flesh. Representations of the ankh symbol, if we can propose it as at least partially a reference to the head, imply the power to see just as much as the idea that rulership comes from the sky above.

Here it is worth noting that prior to the genetic anomaly, the connection between the head and divine right was probably already latent in the culture.

For hundreds of years previous to King Tutankhamun, elongated headdresses of the Egyptian kings and queens functioned as crowns symbolizing an upward connection to spiritual realms. Many Egyptian gods featured such crowns. Made of feathers, reeds, metals and various other elements, they served to enlarge the size of the head and heighten the body. Mary Abram speculates that there are "various symbolisms" accompanying the crowns, generally depicting gods on earth. Objects on the crowns like sun disks and ram's horns may have "symbolized resurrection" in accordance with the Egyptian mythology.[25] Jaromír Málek argues along similar lines: "The king and the queen were the new deity's main officiants, and it was only to them, as representatives of mankind, that the sun-disc extended its arm-like rays."[26]

The puzzle is why the ankh sits in Imhotep's hand. The presence of the ankh seems to stand in a tension with Málek's claim that the god's rays only shined upon the king and queen. But then again, the ankh may have more generally represented the ability to bring life into the world. As noted, the disc on the ankh seems associated with the sun god Ra who alters the seasons; the ankh might then display the warming and setting sun, a recurring symbolic structure in Egyptian symbolism.[27] An alternative but related view is that the ankh symbol is suggestive of male and female anatomy.[28] As fertility symbol, the ankh would not be easily divorced from the power to conceive as well as to perceive. Imhotep, like the pharaohs, would then embody a special connection to divine realms when holding the ankh, which could tie back to the insight needed for his magnificent creative work.

Yet, for Imhotep, the ankh's endowment seems to stretch beyond simply the ability to conjure innovative architecture. He is said to have also exercised the power to heal bodies through knowledge of material things. Imhotep is said to "make limbs healthy" and have "skillful fingers," just as he is also credited with the step design of the pyramids.[29] In later temples, he is depicted with "knowledge and skill closely linked to literacy and scribal craft."[30] Touching stone, leather, paper, and flesh co-inform. Making buildings, crafting bodies of texts, and helping bodies to function well are all entwined. The insight of the heavens envelops numerous material compositions. The world, in this vision, works from common underlying principles.

Riesse argues that "Imhotep came to personify the sage man, learned but also committed to apply his erudition to the welfare of his fellow human beings."[31] In cultivating knowledge about the body, astronomy, alchemy, and architecture, "he was linked [in later scrolls] to the creator of the universe and patron of artisans."[32] Indeed, materiality was a matter of the gods, and the gods desired human betterment as reflection of higher realities.

Thus, the power of the ankh, as one of the most enduring symbols in ancient Egypt, likely references the living more than a head that thinks. In short, the ankh is not displaying the human that rules through rational decision-making. In a mythological society, strident individuality, a high ordering of the lone genius, is nothing if not entirely of and for the gods to determine. If the ankh resembles a head sitting on top of the shape of a body, then it more likely signifies a magical eye inside of the body, an eye of the sun god that sees all only because the spiritual realms stir within and only because the gods spur other-worldly creative powers and forge human connections.

In contemplating the head, we should also not forget that creative mentalizing manifested in the chest or the heart for ancient Egyptians. The minded brain, if we can rethink the name "mind" as a pliable cultural marker here, was broadly an abstract idea: the gift of insight. The head may have been the symbol of divine power, but the mind was in the chest. In some sense, then, the "brain" of the ancient Egyptians was enwrapped by the body, surrounded by the ribcage and invested in feelings of the heart. The importance of the location should not be lost on explorations of mind today. If the brain that sits behind the eyes is insufficient to explain consciousness—as many scholars invested in embodied and "sensorimotor approaches" now argue[33]—then the Egyptians appear ahead of their time. They turn the head and look down toward the body, refusing to describe a head divided from the rest of body. In fact, for the Egyptians, the tongue that speaks and the hand that draws best reflect what they consider insight, or "the mind"; we do not presume that the Egyptians ever detail the functioning of the body as a mind-producing Thing in its own right but, rather, as a phenomena—an insightfulness or visioning—inseparable from spiritual energies. David Silverman explains the over-arching concept:

> The "Memphite Theology" makes a carefully reasoned connection between the processes of "perception" and "annunciation" on the human plane and the creator's use of these processes in creating the world. It ascribes the power behind Atum's evolution to the mind and word of an unnamed creator: "Through the heart and through the tongue, evolution of Atum's image occurred." The word used to describe Atum's "image" is one that normally refers to reliefs, paintings, sculptures, and hieroglyphs (called "divine speech" by the Egyptians). All of these are "images" of an idea, whether pictorial or verbal: in the same way, the world itself is an "image" of the creator's concept.[34]

Although it is difficult to say that "art," as we understand the term today, existed in ancient Egypt,[35] representations were clearly revered and likely because they made apparent an underlying divine creative force; thus, those who made them could be celebrated as spiritually gifted. The artist could be tied to the divine in a foundational way. In brief, the human body was a conjuring or a conduit; the brain had little to do with the moment of inspiration or the hours needed to develop skill.

The historical examination intends to foreground how the generative power of human reason and imagination deriving exclusively from any single, isolated organ remains largely absent from most of human history. In ancient Egypt, as one example, the physical heart was, of course, taken to be extremely important, and it is possible to argue that it was seen as the source of immaterial inventiveness;

however, the qualities of the heart were only conferred from on high. So presenting this as any kind of strict materialist view would be misguided. A focus on the brain organ today, in contrast, presents a very literal connection between materiality and the mind's creative energy, something probably foreign to the ancient Egyptians. In some way, then, we can see how the brain-as-mind is an even greater reduction because the Egyptians required so much more—endowments from cosmological dedications—to annunciate creative action. In other ways, of course, they required so much less by being uninterested in venturing down the long, arduous road of scientific theorizing, experimentation, and replication.

The irony, for us, in taking the brain as the source of creative power is that the organ still lacks the romantic energy of the Thing that we continue to say initiates human life. The first inkling of love, we commonly say, is always in the heart. The beginning of life in the womb is often said to be the sight of the beating heart. The activity or non-activity of the brain is only reserved for death. Take from this what you will, but ideological formations both confer upon the body symbolic configurations and emerge from pre-existing ones.

Today's interest in and engagement with biological structure stands at a long distance from the figure of Imhotep raising the ankh. Nevertheless, the ancient Egyptians look quite like us; they organize life around a definable One, which serves as a locatable core for human creativity. Imhotep's ability to craft the sensational monument from stone—producing an overwhelming "awe" so that imaginations stretched beyond what had ever been thought—compels a narrative about the source of inspiration. If their source was sets of gods, then ours may well be sets of brain findings.

The brain today is a new kind of god. Yet, this is not the lesson of the ancient Egyptians for us today. If we can learn from the ancient Egyptians in a way applicable to today's rush to trace out neuronal pathways in order to understand creativity, then we can say that the lesson is this: our symbolic worlds still influence our material infrastructures, and our bodies are feeling the impacts. As in the time of the ancient Egyptians, what we create changes how we think, and in reverse order, how we think then changes what we will create. Tracing neuronal pathways, in brief, changes us, but it is a method of discovery that looks into the head, which is a one-way looking that stares right past the material generativities of hydrogen-dioxide, mitochondria, and carbon nanotubes. Conditions that we bring about in our landscapes and with our technological capacities inevitably alter our rituals and beliefs. Just as the ancient Egyptians wanted to ensure their empire through temples and ensure their bodies through new medicinal insights, what we do today in our world is not much different; no matter what we do, then, we will not necessarily be able to dislocate our most basic reasons for always seeking to be more and to do more. That may well be something in us that is "of the gods."

Today, the brain alone is often said to judge and guide human experience; accordingly, we pursue the brain's pliability and techno-enhancement. In stark contrast, the beating heart of the ancient Egyptians could do little to make transformative strides unless the gods decided otherwise. The people had no choice but wait for the gods and pray for another Imhotep. Our brain, however, our creator's concept, is right here with us, and it feels—for itself or to itself?—ripe for manipulation. In secular divergence from spiritual suggestions, we now dance rituals around a tightly woven neural circle seeking greater creative power. However, this performance even if inevitable and so different than the ancients may, ultimately, turn out to be equally difficult to dislodge from ruling systems and present ideologies as anything that could be described as more religious.

The brain of Alexandria

Translating the shape of a thing into a social function far pre-dates the Middle Ages. Presumably, the underlying logic is ancient—as old as the thin edge that cuts and the flat stick that smashes. But the full extension of the concept into medical arenas, as seen in the Middle Ages, does not always produce obvious results. When the form-function logic wiggles into reasoning about bodies, a plethora of ideas from various cultural and religious influences get applied to the lines, linings, clots, and blobs. These can clog up open imaginations about how bodies work, what the delineated parts might do, and more generally, how humans *should be*.

Consider an image believed to originate from an anatomist in Alexandria in 300 BC.[36] Here is a dissected, splayed body with a brain floating inside of a crudely drawn round circle for the head.

The depiction is entirely childish to today's viewer. But as Andrew Wickens notes, what intrigues from a neuroscientific perspective is the attention to the blood vessels and the way that the brain is divided into three squiggles. Here "the earliest depiction of the brain" attributes anatomical shapes to functions considered "brainy," i.e. thinking is not being linked to the heart in this drawing, as was often the case in other cultures, as in ancient Greece.[37] The text beside the drawing "attributes the brain with imaginativa, logistica and memoria, thereby showing knowledge of a medieval doctrine known as the cell doctrine,"[38] or the idea that the brain is made up of defined rooms or "cells" that produce functions of mind.[39] Three squiggles, therefore, equals three ventricles that must, so the reasoning goes, embody separate functions.

The loose, cartoony approach to the image should not be criticized too harshly. For one, it exists prior to advancements in the science of anatomy. The image likely served as a general explanation to how human life was enabled by different biological structures. The image may have also aided doctors when learning about connections between blood vessels and organs. It was probably not drawn to detail absolutely everything. Further, the image may have had a social function: to push bureaucrats and religious rulers to reconsider the value of anatomical exploration. It is easy to imagine that the image positioned the anatomist as either an innovative knowledge-seeker or as a rebel or a rule-breaker at a time when touching dead bodies and slicing them to pieces for investigation was patently taboo.[40]

Despite the image's social role, what unsettles, from a representational point of view, is the size, shape, and isolation of the brain. There is an immediate laughability

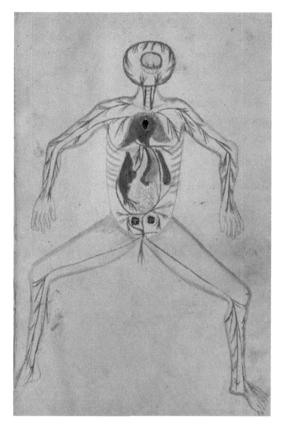

FIGURE 3 "Five figure senses," found in several ancient manuscripts from around AD 1200 but dating to Alexandria 300 BC. Permission granted from Basil University Library, DII 11, fol. 170.

to the small, cartoonish image. The comedy arises, first, from the widely splayed legs and then from the miniaturized depiction of the brain inside of a massive head. The brain is disconnected, much too small, and suspended in empty space. If size is indicative of relative importance, then the heart and lungs make a bold claim on the image; the brain looks to be an afterthought. Perhaps the image exposes the anatomist's ignorance about the importance of the brain or the connections that it has to the body. What, exactly, are those connections?

We might also say that the silly circle for both head and brain, amateurish in appearance to today's sophisticated visual sensibility, suggests little knowledge, or little care, or little attention to detail. There are no glands or nerves, not even a spinal cord—and no indication of a central pineal gland, which would not be unexpected in this time and place.[41] The proposed seriousness still evident here however—a medical image made to coldly document the organization of the body—stands in obvious tension with the crayon-like sketch absent so much detail.

Yet, the image is prior to the development of the scientific revolution and to science practices as we know them today. Moreover, the image precedes a good understanding of how to draw perspective. The artist, for lack of a better word, may not have been able to show much more than what is there. As Wickens explains, "[perspective] was one of the key developments of the Renaissance, and it led some artists to take a new interest in the human form."[42] If realism is entwined with sight and the skill of drawing, then an unforgiving flatness in the Alexandrian image, a body rolled over by a cement truck, requires some forgiveness, and some imagination. The realist view that demands exacting reproduction and careful detail, closer to a set of instructions for an anatomist, is not what we have nor can demand here. This image, rather, looks to be a loose overview.

Perhaps, the anatomist in this time felt deeply like a child in this complicated anatomical endeavor. Considering the flatness and plainness, the body must have seemed, in so many more ways than today, mysterious. From what was the body animated? By blood, by God, or by sheer form and function?

Considering form and function, the most chilling aspect of the image is the isolation of the brain. Descartes would write his "deepest and most lasting legacy" some 1,600 years after the production of this image,[43] yet here in this earliest of medical sketches, the brain is clearly separated from the body and correlated to the imagination, memory, and logic. Intriguingly, the brain floats as if in a glass jar full of water, a tiny vessel out at sea. No spinal cord lifeline to the bodily territories of the arms and legs. No nervous system for calming the ocean of the mind. Brain and body are cut, divided, and distinct.

If Descartes is credited with a turn to the abstract (the mind) as one of the first philosophical attempts at stating what can, through and within a course of reason, be "known" absent the potential trickery of the devil senses, then Descartes' dualist philosophy is not without precedent in medical imaging history. Descartes' reasoning is quite visually present in the Alexandrian image, at least in implication. Descartes' words, stressing the disconnection of the mind, appear precisely aligned with this Alexandrian sketch. As Descartes says, "I have a clear and distinct idea of myself as a thinking and non-extended thing, and I have a clear and distinct idea of body as an extended and non-thinking thing."[44] Descartes can, consequently, conclude that the mind is not the same as the body; thought is abstracted and relegated to a metaphysical world for its origin. The body is alone.

In the ancient Alexandrian image, the isolated brain-organ is credited with holding the imagination, memory, and logic. The brain in the image is divided from a body whose veins, like a spider web, string it together as interconnected substance. Here, in the visual arrangement of bodily matter just as in the creative presentation, we see a historical medical knowledge implying that mentalization and the structures of flesh are extremely difficult to reconcile, at least from the anatomist's point of view. Moreover, if bodily form does indeed intimately relate to function, then the Alexandrian doctors of 300 BC were merely guessing at what brains might do but finding few answers. Staring, then, at the gruesome and imprecise image, we are left to wonder how much revelatory insight meticulous attention to the details of the body's form might be able to yield. We can sit on this thought a little. We can consider it in reference to today's detailed anatomical maps. We might also then wonder if interpreting the body's shapes necessarily inscribes something social, something foreign, onto materiality, something that makes a human into a thing that has, for example, "imagination, memory, and logic" in a brain that floats free.[45]

Scan 3: the brain is a suburban box home

The boundaries of the brain organ seem immediately self-evident—until trying to make it account for the totality of the mind. "The mind," as cognitive neuro-scientist William Uttal notes, is immensely difficult to describe. Because a childhood memory is triggered by a smell, because we have "gut feelings" about our decisions, because we remember things only when we have a pen or a list in our hand, the bounds of the mental can quickly become fuzzy. Philoso-pher Alvin Noë makes a similar observation. He argues that the brain acts *with* the world and, in fact, needs the world to do what it does. The brain is not, in his words, "the sole author," i.e. neuroscience has trouble relating mind and brain because consciousness does not happen inside the brain, per se, but in "a larger system" (6–10). Assuming that he is correct, it becomes possible to argue that neuroscience should look not only to the body to better understand the mind but also to broader environments.

Obscured material elements—the ugliest bacterium, the smallest virus, or the dashing particle of light—may be integral in a complex emergence. Thus, in a context, presuming the insignificance of multiple material forms may be a huge mistake. What if the tiny things turn out to be super active and way too "brainy" to relegate to nothingness?

To be "brainy" in so many currently unimaginable ways is to dethrone the brain from its hoarding of the claim to creative and agentive power. To call the bite that activates the whiptail lizard to asexually reproduce "brainy"—or to see the material actions at play there as forms of braininess, as forms of rudimen-tary "thinking" like responsive initiations—is to reimagine a world jammed full of brains.[46] To re-see the chemical defensive action of plants when in vicinity of the munching sounds of caterpillars as a form of braininess is to uncover the botanical brain spread across the surface of a leaf.[47] Braininess, in this more hospitable concept, need not be exclusive to neurological firings.

Relating mind and brain in any one-to-one way should immediately initiate two questions: the first is whether the brain organ can be so resolutely in the head, and the second is whether any of the many non-brainy things can act brainy enough to spur (or more longitudinally, guide) the production of mental phenomena. Of course, to some extent, cordoning off a brain from the possibil-ity of multiple and extended means of operation is a problem self-generated by the isolated linguistic designation of the organ in the head as "the brain." The identification follows from a much longer philosophical compulsion to see a brain that "thinks" as divided from a body (that lump over there) that is dumb and a world (far out over there) that is diverse and disconnected. But what if this word—brain—was not so strictly confined?

Uttal—who will act herein as a neuroscience representative for the moment—recognizes that many unresolved questions still reside around mental phenomena.[48] He's a realist in that sense. But he still holds out hope for relating the brain organ to all mental phenomena, and he, ultimately, sides with the brain as responsible for the totality of mind. He states, "the mind is neither nothing more nor nothing less than a function of the brain."[49] In this, he speaks for many; but he does not assume neuroscience can do this by itself, within its own disciplinary practices. He realizes the need for ontology, a branch of philosophy concerned with "the nature of existence itself."[50] Questions about what "thinking" or "reasoning" mean in the first place seem immediately philosophical.[51] As he says and advocates, neuroscientists and philosophers must work together if the furious fire of the problem of consciousness will ever be doused.

But I advocate more than partnership and good quality thinking that might jointly uncover how, exactly, the brain does in the end account for the totality of mind. I advocate first tackling a deeper problem, a problem of the idea of a brain, a problem of what it means when things unlike the brain do this or that. I am not out to resolve the problem of consciousness as a mind-brain problem. This book is not about dousing that fire. I focus on the smell of a fire burning within us but see the smoke everywhere. So where is the fire? I often wonder if the heat emanates from something far beyond "the human," a much more basic composition. Like the ancients, I feel this first in the chest, where I detect a precarious feeling about the conditions of our existence, a feeling that hints at something massively misunderstood.

We sweat now over the brain, and we hope that it can explain everything about what we see, feel, and do; but the fire may not be on the top floor and the solution not that confined. The plain truth is that appeals to the brain look too often like a convenience. The brain becomes the pinnacle of a Modernist organ orientation, and even though that sounds like a criticism, that view may help us to deal with an inability to know where to turn and a difficulty discovering something about why and how we love, how and how long we have memories, when and how to see beyond our frames of reference, etc. What we face is an inaccessibility at the heart of materiality. And perhaps as a necessary recourse, we worry about unexpected insurrections in our material conditions. We want to try to halt them. There, the brain, too, looks to be as good of a starting point as any. But can it stand alone? Will it think us out of our trouble?

The cognitive difficulties that follow the ingestion of hard metals, the affective irregularities from specific kinds of bad gut bacteria, the inflammation and stress resulting from toxic air particles cannot be ignored. Questions about bodily permeabilities clang like gongs amid environmental destructions, and the lack of knowledge about bodily processes and their integrations and dependencies frustrates. We need answers. And if finding one, then we will quickly need a good one for that one, until we have just One.

We pray that there can be an end. But, for now, we need a place to go. We need a kind of home or a place where we can start to be in control. We gotta use air and water filters. We gotta have cameras and a security system installed.

If Uttal is correct in saying that "the mind is neither nothing more nor nothing less than a function of the brain," then neuroscience is, in the metaphor of the home, already decorating, making up the bed, and getting ready for the guests.[52] But we have to wonder: is this a kind of premanufactured home with a leaky roof and poor insulation? Will it protect us from the elements? And is it going to feel very cozy?

The brain of Yin and Yang

The origins of Chinese medicine pre-date its written documentation, stretching back far beyond Confucius (551 BC) but developing progressively over many ruling dynasties.[53] The Confucian texts give much consideration to the body but can be read as a socio-political order mapped onto the body—liver, heart, and stomach are the body's ruling team. Even amid such regulation and careful reasoning, no attention to a brain that thinks, guides, or contributes much to healthfulness is found in the early texts of Traditional Chinese Medicine (TCM). TCM texts rarely reference the brain, except as an ancillary problem to be solved in the body's im/balance.[54]

Because a theory of Yin and Yang conceptually guides TCM, balance is central to the healthful life. A body requires symmetrical composing. Divided into upper and lower as well as equal vertical halves, the body feels whole only when no one part lords over the others. Indeed, the concept of symmetry goes all the way down, so to speak, to material things, well beyond the body. In TCM, all things require a proper, balanced form to be functioning bodies. The human body is but an example and, for us, the preeminent one.

Yin–Yang balance in TCM is complex, and the following discussion shows how the concept works once functioning or embodied as a human Being:

> The upper body belongs to yang while the lower body belongs to yin. Other yin yang pairs in the body include the interior (yin) versus the exterior (yang), the front (yin) versus the back (yang), the inside (yin) versus the outside (yang) ... Each organ can also be further divided into yin and yang aspects ... TCM believes health is achieved when yin and yang are in harmony. As already mentioned, the body's physical form belongs to yin while the body's activities or functions belong to yang. Because both the body's

physical form and functions are dynamically balanced, they mutually restrict and depend on one another. The body cannot function if it doesn't have a physical form in which to perform them.[55]

The Yin–Yang concept weaves a logical tie between form and function. Bodily matter is predicated upon an indivisible relationship between a balanced design and healthful function. So what about the brain?

There may be several reasons why the brain and its balanced two-lobed design is largely ignored. One possibility is that the brain does not visibly perform a function that can be easily attributed to its form (without a brain scanner helping to make some correlations, that is). Another possibility is that the divisions of the body (top-bottom, front-back) do not well align with a brain that sits only on top of the body but stretches from the front to back yet appears divided from side to side. A third possibility is that we, from our perspective today, expect the brain to be included because of our own prejudices; we hold an underlying concept of simultaneous chemical and electrical action, but no such concept undergirds TCM. A fourth possibility is that the brain is not historically an object of bodily agency with respect to health because the brain does not digest nor appear to excrete; it does not fuss, except in a headache, which has more obvious causes in alcohol or an unruly family member. And quite possibly, the brain's mushy squiggles appear too wet and gooey, and accordingly, they might be interpreted as too susceptible to bodily juices in the TCM schema. Or, the brain, if understood as a thinking organ, may have simply appeared linked to the body through the spine or, perhaps, seemed to be easily influenced by bile from the stomach or other external factors; that is, the brain could not be of itself. For these reasons, the brain never was a priority for TCM. The brain is ruled by the other organs and lacks agentive power.

Understood as an image of the body, the Yin–Yang symbol has no brain. The happy, clear-headed, good-feeling individual is without stinky urine, free of cold sweats, rosy in complexion but not too hot; the sex is good; playful wit rolls off the tongue. The brain and braininess dissipate into a body whose healthfulness is judged on overall good feeling.

Kaoru Sakatani at the Nihon University School of Medicine argues that a fractal concept for the body in TCM centers around five "zang organs" correlated to five basic elements, which include, "wood, fire, soil, metal, and water."[57] Any disruption of the brain would, therefore, have some underlying connection to an imbalance in one of the zang-organs. As Sakatani notes:

> In modern Western medicine, the brain is the most important organ, acting as a control center. In contrast, the brain is not included in the organs of TCM, i.e., the five zang-organs (heart, liver, spleen, lung, and kidney) and six fu-organs (gallbladder, stomach, small intestine, large intestine, bladder, and triple-warmer). Interestingly, in TCM, the brain functions are scattered over the human body.[58]

FIGURE 4 Traditional Yin–Yang symbol, Creative Commons image. Possibly one conceptual origin of Chinese medicine.[56]

Despite the obvious lack of concern about the organ that garners the most attention today, the idea that the brain is "scattered" indicates an underlying ecological concept for a healthy body. TCM heightens awareness of balance and symmetry, extending those concepts to a relation between inside and outside. In so doing, TCM is not afraid to look beyond the brain to begin to understand the production of mentalizing. In this respect, TCM eclipses many Western concepts.

However, restrictive dualities of thought still guide interpretations of biological processes in the Yin–Yang conceptual schema. Zang organs are predicated upon balance versus imbalance. The search for the all-encompassing One, a simplistic explanation for healthfulness or sickness, repeats itself in the up–down, left–right, form–function orientations of TCM. The One, in this case, is an abstraction manifesting as "balance." Its comprehensibility and tangibility is enabled through the shape of bodily materiality. What TCM has in common, then, with the ancient Egyptians and the Greeks is the propagation of conceptual dualisms as a basis for forging practical solutions to the problems of everyday living.

The philosophical tendencies in common with the Greeks go yet deeper. TCM also reflects a philosophical struggle between using an abstraction to explain all matter, on the one hand, and using material relationships to explain unstable Things, on the other. A debate of this sort between the Greek philosophers Anaximander and Thales ventures into great detail, as explored in future pages.[59] But in fact, TCM rides a savvy line between material relationalism and an abstracted metaphysics. Ultimately, TCM is resolutely materialist yet conceptually mystical. Ontologically, then, TCM constructs an elegant order and simplicity by which to guide and explain human life. In the process, a set of admirable values is foregrounded, values that would not be so easy to generate or uphold if a strong brain concept was driving TCM. If the Western brain concept was part of the picture, then TCM would surely have difficulty thinking the whole body, explaining ecological existence, and seeking bodily improvement through greater degrees of balance.

Da Vinci's brain

The expectation of some readers to see something about Leonardo da Vinci will likely prove too overwhelming for a chapter exploring historical symbolisms of the brain. So I yield, but with dissent. Discussing Leonardo steps beyond the so-called "ancient." And his work, usually framed with giddy proclamations about the revered artist-thinker-inventor, is already over-theorized and over-heroized by humanities scholars. Even so, one image offers a flash of insight into a changing brain concept.

We see in the Leonardo image below an Italian Renaissance brain becoming the Big-B Brain of explanatory power. When mulling over the composition and the notes that Leonardo makes beside it, I think that we find it difficult not to see a belief that hypersensitive attention to the anatomical detail of the head will ultimately enable us to uncover the deepest secrets of our own perception. The sketch in fact makes a kind of epistemological argument, namely, that knowing more is a matter of intense visual scrutiny. Analysis of the image elements brings this forward but also turns us around on Leonardo's suggestion, compelling us to wonder if an epistemology of sight in material investigations is wrong-headed. Seeing may not only be imbued with socio-cultural and disciplinary perspectives, as Leonardo shows us inadvertently here, but may simply be the wrong mode of access for some material entities. To the point: in closely scrutinizing this image, we end up questioning the hopes and dreams that Leonardo seems to have for biological observation.

At first look, the image retains some similarities to ones predating it. Similar to the Alexandrian brain image discussed earlier, for example, Leonardo maps the brain's three ventricle regions directly onto three specific functions—memory, reasoning, and common sense.[61] In this identification, he retains the propensity of early scientific predecessors, seeking a broader theory of mentalizing through the positions and shapes of the brain. However, this image, in both style and detail, is vastly more complex. Indeed, the image is exceptional for its artistic quality and its anatomical correctness compared with previous endeavors.

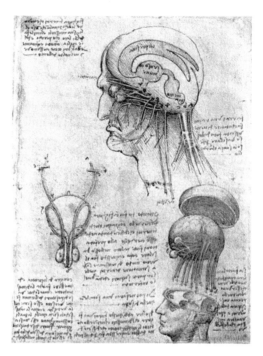

FIGURE 5 Leonardo da Vinci. Composite Rendition of the Brain and Cranial Nerves. Weimar verso (ca. 1508).[60]

With respect to style, shading provides depth, and the invention of perspective is wielded for exacting reproduction of mechanical intricacies. As Ackerman notes, art was resolutely "in science" and fully "of the scientific" at the time; to lack attention to detail would be tantamount to neglecting the values of humanism. In fact, the fame of the Italian Renaissance and its celebration of a free-thinking, self-driven "Man" is inextricably entwined with representational realism, in which Leonardo played no small part. "Leonardo's passion for observation gave him the capacity to challenge the dicta of artificial perspective, which were strictly geometric and abstract, unrelated to perception."[62] He aimed for a truth of human matter, how the thing "actually worked."[63] A seemingly endless series of revelations about anatomical arrangements and biological functionality amid the grand wonder of consciousness could not but encourage—if we put ourselves in that distinct time where Judeo-Christianity and science were co-informing as well as clashing— strong feelings of human magnificence. Exceptionalism was the word, and human exceptionalism above all.

Leonardo makes room, then, and not surprisingly, for the sensus communis (the proposed location of the human soul) in his anatomical sketches. Yet, given the cultural atmosphere, he deserves praise—despite my own earlier grumblings about the exuberant praises showered on him—for his desire to observe the whole organism and push beyond simplicities. Indeed, the sensus communis note may well have been a cultural requirement; whether Leonardo merely accommodated such strictures is not known. Whatever the case, his desire to expose matter's link to other forms of life and to deduce the production of *life itself* through analyzing the activities of the whole body seems especially cutting-edge, even today. He was clearly an experimentalist developing a theory, and he seems willing to start everywhere and try anything.

For example, in one famous story, Leonardo punches a hole in a frog's spinal cord. He states: "The frog instantly dies when its spinal medulla [medulla oblongata] is perforated. And previously it lived without head, without heart or any interior organs, or intestines or skin. Here therefore, it appears, lies the foundation of movement and life."[64] Appreciating the role of the nerves, bones, and muscles, Leonardo's avid exploration of the many links from the brain down to the body is visible in his inky scrutiny of flesh. Although the head floats on the page of several of his drawings, the framing could be pragmatic only; the octopus-like arms reaching toward the body suggest an understanding that the brain is not an independent organ and that perception and bodily action are linked.

Leonardo's special attention to the structure and function of the eyes seems to also foreground, or referentially imply, an obsession with looking closely. The eyes and eye muscles in the sketch, in fact, appear to be of prime concern; if not reading all of his notes, which address various issues, we could almost come to believe from the drawing that the eyes are the main function of the brain. To see, we might want to believe when encountering Leonardo as the artist as genius, *is* to think.

But his sketch of the human head, often called "Composite Rendition of Brain" or "Study of Brain Physiology," leaves open a question about how much can be learned from looking at the brain, no matter the minutia of detail. In fact, what Leonardo teaches here is something of an opposite and unintended lesson: that concepts of the brain are bound so inextricably to social values, symbolic associations, and cultural narratives that looking at matter is not sufficient to explicate the function of it. This is true, of course, for brains. Despite drawing a fastidious image, Leonardo demonstrates an astonishing lack of sight.

Examining several da Vinci drawings, Ackerman concludes that "major aspects of the drawings illustrate points determined theoretically rather than empirically."[65] Quite simply, da Vinci did not draw directly from sight; the hand was guided more by the mind's eye than the eye. As Ackerman says,

> The grid of lines, for example, is intended to illustrate the conformity of the head to a system of privileged geometrical proportions. Further, the drawing illustrates the medieval doctrine that the vertical and horizontal axis of the skull must cross at the site of the sensus communis, or common sense, where all perceptions—of sight, sound, touch, and so on—were believed to be gathered; this was considered the seat of the soul. According to Plato and Hippocrates, whom Leonardo quotes, the soul must activate the entire body and, in particular, must transmit seeds for reproduction from the brain to the genitals. Accordingly, a channel must be provided

through the marrow. The top of the spine in this drawing has a large interior channel, which Leonardo would not have found in his skeleton.[66]

Despite imaginative additions, Jonathon Pevsner notes that Leonardo's drawings do improve with time. The one featured here in this book is a later sketch. Yet even in showing improvement, Leonardo still crafts mechanistic and slightly amusing interpretations specific to Italian sensibilities. As Pevsner explains:

> Leonardo invoked a military metaphor to explain how motor output is also controlled by the senso comune and the soul. As he put it, "The nerves with their muscles obey the tendons as soldiers obey the officers, and the tendons obey the senso comune as the officers obey the general. Thus, the joint of the bones obeys the nerve, and the nerve the muscle, and the muscle the tendon, and the tendon the senso comune. And the senso comune is the seat of the soul, and memory is its ammunition."[67]

Leonardo makes evident how representations of the body stand as time-dependent emergences forged among numerous actors and interacting human and nonhuman dependencies, or what we might more astutely call agencies. The tools he used to dissect bodies, the eyewear that he saw through, the institutional structures through which he could or could not get access to bodies (which determined how fresh the cadavers would be), the ink and paper employed to note the detail, and the people, of course, peering anxiously over his shoulder seeking religious or political affirmation in his findings forged a network of relational power. These, together, we can see quite clearly now, mesh to make his power of vision.

Scientific practice, like other practices, remains subject to socially acceptable formations, the right protocols of experimentation, the relevant institutional agreements, the means of access, the financial availabilities, etc. Bruno Latour says it best:

> modern societies cannot be described without recognizing them as having a fibrous, thread-like, wiry, stringy, ropy, capillary character that is never captured by the notions of levels, layers, territories, spheres, categories, structure, systems. It aims at explaining the effects accounted for by those traditional words without having to buy the ontology, topology and politics that goes with them.[68]

Put another way, the drive to capture and solidify parts overwhelms the idea that the world is intermingling and emergent, not so discrete, no so easily upheld as separated entities out on their own or "thingly," and not easily dissected. Latour would never presume material entities, like bodies, to be so simple as to be self-evident or self-sufficient. Likewise, he could not see scientific action as disconnected from local affordances and constraints.

Taking up Latour's view, we must say, as he does, that "instead of starting from universal laws—social or natural" when trying to understand knowledge formations, we should, rather, begin from the localized practices and beliefs.[69] Turning this idea back on Leonardo might seem to condemn his project, at least to some extent. But then again, Latour does not eliminate the power of science to develop conclusions or re-create the outcomes of experimental set-ups. Latour does not oppose a real world out there. Latour, rather, suggests that local geographies and the special arrangements of things in a time and place make all the difference to the exploration, to what can be accessed, and to what can be restaged and validated. Consequently, how often, and to what extent, observation yields to prior interpretation or when exactly socialization reconfigures matter is a question of who, what, when, where, and how. Leonardo himself seemed unable to escape the background of his cultural knowledge, the limitations of his tools, the politics of his day, and the allure of his own philosophical vision.

We can say more, go further. In charting sight, Leonardo maps veins, tubes, and linings, but he never realized just how far away he was from the mechanisms of sight. That is, if the image he crafts persuades us to invest in the power of observation to uncover material formations and their processes, then when enamored by it, we forget that material mechanisms can operate beyond and out of sight. They can be invisible to the eyes, as in momentary exchanges of oxygen or electric activity, but they can also prove to be invisible to all of our senses, even to our machines. Donald Hoffman, a neuroscientific expert in visual illusions and the philosophy of mind, argues that all forms of human perception must also be taken as subject to evolutionary fitness; the conclusion in his view is that external material realities may not always be what they seem and that "accurate" perceptions are more likely to be those that provided fitness benefits. Pushing the idea to its more radical conclusion, Hoffman even suggests that humans may never have been able to access "the objective world" through sense perceptions. Instead, the features assigned to the world, he argues, might simply reflect the structures of the human brain and its processing capacities, not the truth of the world itself.[70] So looking back at the Leonardo image, we can see the cultural inculcation of a suspicious epistemological notion about sight. We need not adopt Hoffman's view here to say that knowing is a tricky thing, and any claim that we speak need not hold the presumption of a constant and strict correlation between our human sight and an outside existing precisely as-sighted. This is the case even if we feel that we are able to see more with Leonardo and choose to celebrate his championing of the careful investigator as we preserve the adventure that is scientific discovery.

Despite the various aforementioned "problems" with sight, we can nevertheless try to instill hope of propping up the progress narrative for scientific observation: looking closely at the shape of human biology does, over time, appear to lead to better agreement across time and cases, or what we might call representational improvements. We can assuage our epistemological fears with this and say that better vision advances medical understanding. Surgeries and medical alterations to bodies can and do improve lives. Leonardo's effort to detail the body invests in this positive orientation and, thus, aims at describing the universal and the irreversible of the human body. We might then call Leonardo's later sketches "improvements" precisely because we still agree on the anatomical parts and connections and because our tools still confirm them and because they work for us.

Whether or not we hope to one day achieve a perfect, objective understanding of the body's mechanisms, Leonardo displays how we must care for the ways that the body gets infused with the body politic. Flesh is segmented and disciplined in representations, which order interventions. Images of the brain, more than any other area of the body, guide self-understanding and contribute to a contemporary arrangement of Being. Yet too few are the times that we choose to see the body's porous exchanges, even when we can, and many exchanges, have no doubt, extend well beyond the capacities of representation, especially as the organ takes up a majority of the frame. But much more of entwined matter remains. Indeed, no artist could draw it all. No artist could draw so much. No neuroscientist could ever see so well.

Representation's trick—the making of and the assurance of a Thing inscribed—must be exchanged for generative Becoming, which is the assurance of being wrong, of complete inadequacy, of constant iterability. But Becoming's trick, namely, the reverse assurance, the assurance of change and the constancy of poor vision, might be equally damaging at the particular time when a performance should be understood as a routine. So we face a bind. What can we do? If we turn to Leonardo, then we can at least suspect that whatever we do, we will always be showcasing our limited vision and our era's dreams.

The psychic's brain

A mystical rhetoric of secret power inside of the human head circles back around to the image of the pinecone and the pineal gland. Here we encounter the psychic's image of a magical brain, which is both old and new. And we are able to see what the contemporary imagination can conjure when mining ancient neuro-symbolics.

The pineal gland is low-hanging fruit in an orchard of mystical symbols. A quick online search reveals the extent to which the pineal gland is invoked by psychics selling spiritual healing, secret insight, and colossal aptitudes. "Gifted medium, healer, and intuitive," Andye Murphy exemplifies the rhetoric. An article on her blog site states,

> The third eye is our greatest gift to connect us to source and remind us of a universe much more mystical than that which we perceive with our physical senses. It's through the awakening of the third eye and its corresponding pineal gland that we're able to attain supernatural feats of telepathy, psychic vision and an intimate connection with God … Because it's inside the brain but outside the mind, a paradox is formed, creating a loophole to escape the contradiction of this reality. The Egyptians taught that when we can contain this paradox, freedom is achieved. Every Mystery School taught the metaphysical ways of being: how to be awake within the dream, to walk between realities and surpass the limitations of humanity.[71]

The passage demonstrates the allure of the brain for such psychics. There is a seductiveness when the brain's materiality is attributed to mind, which has an inherently mysterious quality; the inability to account for experience or to fully explain how consciousness arises, even neuroscientists must admit, makes the brain a sexy symbolic. However, the discourse presented in the passage persists relatively independently of the contemporary neurosciences. The passage more so seems to evidence the historical longevity of a struggle to overcome the Descartian dualism of mind-brain.

Presumed divisions between mind and brain point to a metaphysics where the brain as an organ of the body is tied to earth, denigrated and lowly, unable to be a conscious mind or to see the so-called higher realms of intuitive awareness (or in some cases have super powers) without additional, exterior and special assistance not of the earth. The psychic predicates an ability to "surpass the limitations of humanity" on this dualism. Indeed, in the passage, the true capacities of the mind appear chained down by the brain, living within an unfulfilled state of bodily submission, even as the mind desires elevation and freedom. This is an old rhetoric. It is the dream of a Platonic release from the cave where men sit unable to see "the real" world outside. There, the sun shines brighter. It is also a Descartian re-visitation where heavy flesh and bones do not participate in a beatific vision, which is relegated to the lighter, airy thinking "I am" of the mind alone. In those adjectives (heavy/light and airy), earthly symbolics help to co-construct the underlying concept; to be a mind and think is to be free, like a bird, whereas to be a body is to be heavy and weighted down like a stone.

The irony for the psychic discourse is that Descartes, as an early rationalist, still seems to shape the discourse even though the psychic has a stated aim to escape what can be rationalized. The psychic claims to want to escape technical, Western, scientific logics that deny an original "source" and cannot access the "mystical universe," yet adheres to a cosmological logic consecrated by Western rationalism. The pineal gland negotiates the impasse.

Adhering to dichotomous logics while wanting to reject them can be rhetorically traversed by the gland, once positioned at a crossroads of the material and the magical. The gland, so we are told, elides dichotomies, namely, mind–body, source–earth, higher–lower, emotive–rational. The cosmological construct being used here surely is a historical product; however, the choice might also indicate something about aesthetics. Put another way, we can imagine how waking suddenly under stars that spin in a circle and watching the rotation of the sun or the cycles of the moon might inform a symbolic reading of the delicate turning appearance of the pineal gland. Even without the rhetorical seduction of ancient imagery where Egyptians and Babylonians are seen cradling similarly pineal gland-like objects— including pinecones—the shape may be understood as a fitting key to the mystery of being human amid the grand Milky Way.

The seduction of the psychic's brain—that is, the allure of a kind of brain being imagined by psychics such as Murphy—is not constructed solely by an underlying adherence to moves inaugurated in a Western tradition of philosophy;

this brain is also made from a bodily sensation. We feel the force of the pinecone when seeing the spiral repeated in the fields and forests. The key to enlightenment, amid our seemingly in-built attractions to repetition and symmetry, surprise and novelty, could only be reasonably inside of this perfectly spiral object, perhaps we are more willing to believe. Once the pinecone shaped thing is found inside of the brain organ—the very part of us designated as the thinking thing!—arguing that the pineal gland offers a way out of the mind–body dichotomy makes sense, or at least makes more sense than, say, arguing that the metatarsal bone does this work.

Appeals to the ancient imagery of the pinecone may also function persuasively because the social and political realities of the original creators that designated the pinecone as important sit just far enough out of reach to not be fully known, and hence, to instill a mystery. The pineal gland can be positioned as a secret long lost about amazingly deep knowledges. The lack of situated historical knowledge makes it easier to claim today that something has been shoved aside by Modern humanism's scientific culpabilities. Looking for something from the past to provide guidance might also play into the notion that societies get worse by the decade, cooperating with a sentimentalization for previous social arrangements. Ancient peoples would, given the narrative of progressive degradation, be more "in touch" with the essences of nature that connect to higher realms and original sources.

The appeal to the pineal gland for greater human power, however, forges a strange paradox. Mental power arising from bodily materiality seems staunchly counterintuitive when the body is also simultaneously situated as hapless flesh. Overcoming a mind–body/natural–supernatural dichotomy through awareness of the pineal gland's spiritual force uses the body as the means to access mentalizing powers. Even as the body limits the mind's abilities, the psychic suggests, so can secret practices unleash hidden synergies of the mind–body. This is a strange paradox—crafting a bodily means to escape a prison forged by the body—and it seems irrevocably Modern, insofar as it recalls what Katherine Hayles describes as the old "liberal humanist" worldview.

For Hayles, liberal humanism indicates, among other things, that the mind is imagined as potentially free and "the body is understood as an object for control and mastery rather than an as an intrinsic part of the self."[72] The Modern dream, then, is to re-technologize the body to forge new compatibilities with a mind that craves release and can achieve supernatural things if given the opportunity. Hayles' alternate dream, in contrast to the technophile, the transhumanist, and in this case, the psychic, is choosing *not* to be "seduced by fantasies of unlimited power and disembodied immortality that recognizes and celebrates the finitude as a condition of human being" (5). To seek another supernatural ability is to turn away from an embrace of the limitations of the human. Love thyself, namely, love thy body, Hayles seems to be saying.

Hayles' words rebuke cybernetic and futurist discourses about transcending the body through computer technologies, which recycle "mutation and transformation as a central thematic for bodies."[73] If psychic discourses are any different, then the power of release exists already inside of the body. For the psychic, the body acts as a kind of hapless prison guard over the mind/soul; or maybe it is more like the body and mind play out a sit-com where the girl and the boy living across the hall from each other are best friends trying to find the perfect mate, while the audience knows for seventy-five grueling episodes that the best answer is right in front of them: they should be together.

The emphasis on the smallness and unrecognizability of the pineal gland probably strengthens the rhetorical case for the psychic. The peanut-sized organelle becomes the means to transcend the body's massive weight and earthly stupidity. Extraordinarily salacious, isn't it? Discovering unlimited power in the smallest, hidden thing seems, on the face of it, wildly outrageous. But rhetorically speaking, smallness opposes the obvious. Smallness thus helps to explain why so few people have achieved any transcendence. In addition, the individual, once positioned as a seeker out to know the Truth of the universe, can experience a sense of discovery when uncovering the small and unnoticed thing. In realizing the secret, the seeker not only has a reason why others have ignored it but also, crucially, obtains superiority in finding it.

Finding great power within, and not from an exterior endowment, lends a special sense of superiority; rejecting the obvious power of majestic kings or decorated priests turns seekers toward their own inner intuitions. Consulting one's own phenomenological experience to determine who or what is right about essences or "the nature" of things rejects hierarchical social orders. Psychic discourses, in this way, appeal to the rightness of one's own feeling in exactly the places where the rhetoric falls short of credibility and evidence from an outside. In not having the credibility afforded by scientific accreditation as, for instance, a medical doctor might have, the psychic asks audience members to instead consult themselves. Likewise, in not having the ordered logic of, say, a philosopher outlining definitional propositions, the psychic asks audience members to consider whether they, regardless of what anyone else says, think that supernatural powers are possible. Self-reflection on embodied states is intimate to the success of proposing unlimited access to the fantastical.

There is still yet another rhetorical maneuver in the call to the pineal gland's enlightenment: the appeal to pragmatic outcomes. To comprehend the unknown of the body is to make the body able to function in society in new ways, namely, to better control social realms. For example, by foregrounding what "Every Mystery School" has sought to achieve—a way to surpass the "limitations of humanity"[74]— the given presumption is that there is room to conjure secret powers over the world inside one's own folds of flesh. Opening the senses makes a usable body, one able to *do something more*. What the seeker might do, other than pay the bills of the psychic, remains unclear. But that seekers can then feel that they can grow their abilities— see more, hear more, and gain hidden insights about family and friends—are all pragmatic motives.

At this point, we might return to popular discourses about the neurosciences to consider commonalities. Thinking about the psychic's appeal to the pineal gland allows us to consider what The Brain of pop culture offers. Without question, the brain is positioned as the most relevant pathway for extending human self-understanding and perhaps even for obtaining new capabilities in service of an innovation economy dependent upon cultivating specialized (more, ever more) intellectual skills. If materiality is charted with exactness to improve functions, then neuroscience, once applied, also aims to expand the power of the brain as a way to rejuvenate the mind; pragmatically, we believe, like the psychic's followers, that we will be much better off. As Thornton notes, brain training manuals and self-help books continually "describe the brain as the cause of all our success and failures."[75] We should "improve ourselves and our societies" with brain science. Yet there is a catch. The brain is "deployed to support claims to individual agency and responsibility," but the outcome is really up to you. As Thornton says, "You must take specific actions to guarantee the health of your brain."[76] The same is true for psychic's appeals, which promise new life but which can be phrased to blame the dissatisfied individual for not doing enough, for being too weak or too lazy to discover superpowers.

Of course, there is an important difference between the psychic and those working in the sciences of the brain: most obviously, the brain sciences require material evidence for claims and seek replication to step beyond an individual's politicizations. The same cannot be said for many psychics. Likewise, those in the brain sciences move in measured ways to avoid engaging in exploitation, whereas psychics are a gamble, not always under regulation in this respect. But those differences should be self-evident. What proves more compelling are the philosophical dreams and the emotional compulsions shared in the desire to see more and to know more.

Any impetus to create a mental transcendence, to push humanity to new heights, as is often touted in the transhumanist movement,[77] aligns contemporary technicity with ancient rhetorics of mystical power. Any hope for a "higher" plain of existence reconfirms the suspicion that "we have yet to catch up to Nietzsche," as LeMothe puts it when decrying the disavowal of our own sense capacities in search for yet more than the often frail, human body as encountered.[78] It may well be a contested claim to say that the psychic has anything in common with those who dream that brain science will radically expand human capacities, but it seems uncontentious to say that we do not want to accept our present limitations and submit to our vulnerabilities.

We feel the powerlessness of our bodies already in the process of dying. We must "not go gentle into that goodnight."[79] If we search for ways to last longer, we cannot be blamed. Death is the enemy, and history is a prophet screaming that we will lose. But what choice do we have but fight?

The disavowal of dying flesh and the embrace of the imagination about the capacities of the mind may be a perfectly sensible response to standing face to face with a vast unknown. We do not know much of the material world, except that things crumble over time, except that the nearest star is 4.22 light years away (or $2.4808e+13$ miles), except that more than three-quarters of the universe is dark energy and dark matter.[80] And we are not sure about what aspects of things we do not know, nor how to know them.

We wonder if the body is enough to know what we want to know about the world, much less to know enough to keep on living and improve our condition. "The body exceeds our knowledge of it," Spinoza remind us.[81] Thus, it only seems fair to recognize that in struggling to give shape to the world, what the "Mystery Schools" aim to do shares a common exigence with what many of the deepest thinkers aim to do, i.e. forge new narratives about materiality, rewrite the body, and achieve more compelling personal and social ends in the construction of wild imaginations.

We should not bristle too much at comparisons. The philosophical effort to think what has not been thought, to feel what could never previously be imagined, and to reach across the abyss, to create, and to live inside of our fascinations, is a shared and wonderful commonality. We might, rather, bristle at trying to surpass the bounds of flesh as means to achieve dominance over others. In fact, if there is any obvious indication that psychic rhetorics remain all too human, then it is in this predictable move: trying to tap the great energy of a hidden beyond to secure one's own will and "surpass the limitations of humanity."[82]

Scan 4: neuroscience secretly adores Anaximander

Seeking a material unity, an ultimate simplicity in matter alone, is a relatively recent phenomenon in human history. Gross credits Thales of Miletus, 624–546 BC, with initiating this intellectual exercise in the Western world. "Thales' cosmology was fundamentally different from the Babylonian and other pre-Socratic ones for two reasons. First, he left gods such as Marduk out of his schemes. Second, he sought a common element underlying all phenomenon."[83] The common element for Thales was water, "the source of motion," i.e. the essence of living being, the only "Unbounded" substance.[84] Water made for a good choice because it took no defined shape, was essential to life, and had commonalities with the moisture of human blood and sperm. These differences staged water as both essential and flexible, allowing it to fill a role as the unifying substance of all matter.[85]

Although the ancient Egyptians also associated water with life in revering the Nile, for example, Thales seems unique in looking to nature for the essential, "primordial element." He is often credited with being the first "scientist" because he develops a hypothesis from nature. But this designation proves apropos also because he establishes "a unitary principle of explanation" that can be applied to biological materiality and, in turn, shed light on the behavioral.[86]

Thales' project sounds familiar, of course. And the philosophical critiques of Thales' view, as delivered to us through Aristotle, probably will as well. They mirror our ontological debates about what matter *is* and how Things come into Being. As a result, they help us to see how contemporary ideas about the body relate to very old ones and how our methods for studying the body rely inherently upon what we imagine the body to be.

In ancient Greece, Thales becomes engaged in a battle about the nature of Things and whether they retain essential qualities independent of each other or whether they take form from each other like mutating cells splitting and changing along the way. Whether all Things ultimately derive from some abstract or metaphysical source or whether they come out of interrelations of material earth is the attendant dilemma. By way of a quick summary: the former view proposed that Things look to humans like objects because they derive from or they mirror an inaccessible or spiritual realm that lends them their shape; we can follow contemporary philosopher Graham Harman here and call this view "undermining." In contrast, the latter view, proposed by Thales, suggests that Things emerge from splitting and meshing, or from material relations alone, which Harman calls "overmining."[87] One is material-ist and the other, if not spiritualist, is at least working with abstractions. In rebutting Thales, a contemporary philosopher named Anaximander initiates this two-sided lineage.

Anaximander argues that water could not possibly be "Boundless," as Thales suggests, because not all material Things have been subsumed by water and its characteristics. Not all Things are wet or slippery. Conse-quently, those basic Things that humans encounter, such as water and fire, must have been formed by an inaccessible or yet undiscovered Boundless Thing. Aristotle records the debate as follows: "They [air, water, and fire] are in opposition to one another—air is cold, water is moist, and fire is hot—and, therefore, if any one of them were unbounded, the rest would have ceased to be by this time. Accordingly, they [Anaximander and company] say that what is unbounded is something other than the elements, and from it the elements arise."[88]

The debate centers on whether there is a single explanatory material sub-stance on earth or not. Anaximander's position is that this might be possible but probably what is needed is abstracting the original "One," setting it apart from the elements. At first glance, the move looks a lot like undermining, a turn to the metaphysical or perhaps the inaccessible, although the exact detail of Anaximander's position is not immediately evident. But Aristotle presents an alternative solution. He suggests that water "was internally undifferentiated," meaning that all Things might "separate out" from water, while water, as the original Thing, can be deduced to be primary because it remains the only Thing apparently absent a shape.[89] Aristotle's logic reads as an attempt to rescue Thales' relational interconnectedness of matter.

More recent and creative proposals still mirror old ones. Graham Harman's Object-Oriented Ontology (OOO), for instance, advances the idea that Things need not be relationally emergent nor retain some metaphysical quality to make them a specific X Thing. OOO seeks a middle ground by arguing for an essential, realist aspect of an Object that cannot be accessed by other Objects (a "withdrawn" part), which, therefore, makes each Object its own Object.[90] Despite the creativity of the proposal, we still end up with a basic dialogue: one where we argue about whether Things entail an essential element or not, what that might be, and how this can come to pass.

A foundational split in how the world *really is* stutters on repeat. We are caught in a loop of a Green Day song, "Shit's so deep you can't walk away ... I beg to differ on the contrary ... Shit's so deep you can't walk away ... I beg to differ ..."[91] The radio station keeps playing the tune because it is such a big hit.

The centrality of sense perception in the debate underscores why Thales is considered a scientist. Stated in a way more to the point: the intimacy between geometry, or shape of Things, as what differentiates Things (or shapelessness as what makes a Thing able to birth other Things) is intriguing because it suggests that, from the earliest days of the ancient Greeks, deciding on the status of matter was a matter of sight or, broadly, of aesthetics. Seeing that a Thing could visually or physically appear to change form offered an answer to the question at the heart of the debate, i.e. how Things come to Be. Yet, equally as interesting, an appeal to qualities/characteristics that humans can feel (like wetness) aided sides in the debate. In this, there is a lesson to be had: humans have tried to think shape and qualities together for a long time to answer questions about materiality (or the "nature" of things), and two answers seem immediately viable: relational Becoming or abstraction, which can also be phrased as philosophers siding more with change or more with permanence.

Just as with philosophers today, Anaximander and Thales both have good reasons for rejecting the other's proposal. For Anaximander, we imagine, the issue of origins for water was too perplexing within a monist concept. Or, perhaps the historical reliance on the gods was too socially powerful for all matter to be relegated back to matter. Or, perhaps the confines of what he took to be rational thought did not allow a conclusion where the qualities of water (namely, wetness) could not be extended to all things even though the shape of water could, theoretically, be molded to mirror all conceivable geometries. However, for Thales, Anaximander's proposal of an abstraction seems to be missing the point of the question. The point was to be material-ist. Perhaps appeals to an unknown or not-yet-discovered source of all things felt like another magic trick. Whatever the case, despite being differ-ent answers, overmining and undermining reflect a broader imaginative desire, as Gross says when speaking of Thales, for "a common element underlying all phenomenon."[92] One way for all things to become a thing.

Here we can turn back to a brain. In doing so, we note immediately that neuroscience certainly does not attempt to explain everything. Yet, setting neuroscience in reference to this ancient debate brings to the fore the fact that neuroscience serves as the most authoritative, often most pertinent, con-temporary basis for explaining perception. As Schultz notes, "the definitive way of explaining human psychological experience is by reference to the brain and its activity."[93] If "for all intents and purposes, we are our brains," as Reiner puts it when explaining the typical neuro-centric view,[94] then the brain sciences command the authority to declare what is "actually true" and "really real" and what is not. Accordingly, interpretations of brain data that the neurosciences tout are likely to have implications extending out to our perceptions and, thus, potentially our understanding of all things.

Rehearsing the Thales–Anaximander debate is not without some pay-off, I hope. We can try to see how it could inform us today: crucially, the question of the brain (and how it functions or what it does, exactly) cannot be answered except by toggling back and forth between the same two features that help decide the sides of Thales–Anaximander debate. On one side, we have qualities (i.e. our mental life—a tender thought or a harsh feeling), and on the other, we have shapes of material things (i.e. the neurobiological or neuronal complexes and connections). Here again, we return to the abstract and the concrete, felt intangible kinds of qualities and tangible material geometries.

But this is tricky business. Geometric explanations contend with material complexity and constant alteration; that is, to draw on Anaximander's viewpoint, central qualities of seemingly relevant geometries could well not be very apparent, always incomplete, partially displaced to other abstracted worlds (or always in a "withdrawn" state, to use Harman's conception here).[95] If we cannot connect all mental qualities to the brain—just as Anaximander could not find a way to connect wetness to the presence of fire—then how could we ever feel good about relegating abstractions to the realm of the impossible or to non-consideration?

We might simply respond that we feel good about ignoring Anaximander's abstraction because neuroscience seeks conclusions through the concrete: through machines, through collectable data, and through scientists coming into an agreement that stands above any wild theories or individual idiosyncrasies. So abstractions are useful—but only if testable. Indeed, Thales proposes something like this. He does not entertain any of Anaximander's suggestions unless they are material already. In this, Thales seems to be the closer correlative of the neuroscientist. What goes unresolved, however, in that way of proceeding through the debate is a question of whether repetition of experimental outcomes is derived, in the first place, from the upholding of another, underlying abstraction. Anaximander could well assert that Thales arranges the world in such a way as to not see the possibility or need for an abstraction, and this happens precisely because Thales adopts previously another kind of abstraction—an ideal of perfect and/or accessible material interdependence. "Can Thales," Anaximander might whisper to us, "be so bold as to assert that he knows that the world *really is* absent an abstracted unity or source and yet adopt his own underlying concepts, which function a lot like the abstractions that I (meaning Anaximander) propose?" We should remember here that Anaximander did not propose a supernatural abstraction, per se, but it remains unclear what kind of abstraction he suggests; yet, whether Anaximander suggests a new "abstract" material or a supernatural source, he could nevertheless advance the above argument. And if he did—and we imagine that he did—then we see in that argument a reflection of philosophy today and neuroscience, both grappling with materiality.

Confronting these kind of intractable dilemmas returns us again to the original Thales–Anaximander question: is there a common element underlying all phenomena, a principle, force, or mode, that can account for the vast diversity of Things? And if so, can we detect it? And if so, can we then outline this "common" without adding something foreign?

Here the brain is crucial. The brain is like a gangster that tells us everything without telling us whether it is a liar or not. But it tells us what we need to know, or at least we hope that it does a good enough job. We are breathing, after all. We ask others to see if they agree—those are other gangsters, we know, who are in on the scheme, but what else can we do but ask them too? We hope that these little smucky brains give us something good, some meager access to the world. We wonder. We doubt.

So maybe we work to crack the crime organization from the inside out. We look inside the brain with special tools. We are like the FBI. We listen in. And the fact that we must use our brain to analyze it is something like recruiting an insider to tell us more; no, that is what phenomenology does, but whatever the case, we have played out the metaphor to make the point.

The tools of the neurosciences, we hope, tell us more about our perceptions and find a way out of the ontological labyrinth. Nikolas Rose makes this hope explicit, stating that "the most durable philosophies of the human have always had a very close relation to contemporary medical and scientific practices" and new brain findings may now "intuit a new ontology."[96] His stance is rooted in the foundational belief that neuroscience will eventually instill a new ecological perspective on human formation. This celebration of neuroscience is also a reassertion of Thales' overmining, as he suggests that insight about the brain will ultimately show the human to be composed from shifting sets of material relations. As part of the proposal, Rose argues that neuroscience can "move beyond reductionism as an explanatory tool, to address questions of complexity and emergence, and to locate neural processes firmly in the dimensions of time, development, and transactions of the milieu."[97] Putting aside his dedication to relationalisms (what we understand as Thales' position), this enthusiasm for neuroscience and trust that its experimental methods can exceed their structural limitations to offer a satisfying commentary on interconnections with our environmental "milieu," seems more than a little optimistic.

What Rose overlooks is not only the limitations of neuroscientific perception to arbitrate, on a philosophical level, the debate at hand but also the "terministic screen"[98] of neuroscience. The neurosciences develop conclusions from and through ideas about the brain forged from adoration of computer discourses, mind–body divisions, obsession with normality versus abnormality, limited by its technical approaches and disciplined designs. It seems unlikely, then, that anew ontology will arise, or that any ontology going far beyond the organ will, at present, be able to rush out of the neuroscience lab.

Continually foregrounding neurobiological processes might eventually over-turn a Western socio-cultural obsession with the self-made, agentive human if neuroscience can expose interrelations with food, microorganisms, climates, etc. But this seems beyond the neurosciences at this point. Equally, and paradoxic-ally, putting the emphasis on the brain might reify the traditional notion of inde-pendent and free "Man." For instance, when Coulter argues that neuroscience confuses its own study with socially salient "metaphorical constructions" that are "radically misleading," he does so in order to emphasize the field's deep enmeshment with old ideas about human social life and with the dreams of systematization and governance based on neural delineations of capability and normality.[99] And this is a problem, as he adamantly points out, because "brains do not receive, transform, or process 'information,' they form no 'representa-tions,' and they do not 'store.'"[100] There are too many "purely conjectural attri-butions to brains" built from cognitivist theorizing while "linguistic and social" components are extracted.[101] Similarly, neuroscience's inability to escape social categories have led van Ommen and van Deventer to note that "the neuro-chemical self" or the ceaseless self-monitoring of the body,[102] displays striking "similarity to the neoliberal subject"—precisely because economics and neurosci-ence are not swirling in separate spheres.[103] Put simply, neuroscience more often cooperates with than combats discourses promising self-understanding, control over the body, prosperity and personal empowerment, and traditional social and psychological stability.[104]

Perhaps what neuroscience most clearly offers is a way to view our own assumptions about reasoning. Neuroscience may not be able to solve philo-sophical questions on which it, as a practice, depends; however, the sciences of the brain—simply in being practiced in a kind of way—give insight into how we are prone to think about the body and the world today. Thus, what neuroscience offers is awareness of how many different Things—like scanning machines, magnets, algorithms, ways of evaluating a participant's suitability for an experiment—are arranged to make sense. But those arrangements also intensify the underlying questions.

When held to a strict focus on neurons, neuroscience might appear aligned with Thales' investment in a complexity of relations, but most humanists and social scientists studying neuroscience would likely suggest that the field more often appears enamored with Anaximander's totalizing abstraction. The ontol-ogy that Rose envisions, a concept of the human entwined, formed over time, and able to change with its ecologies, is a beautiful one, akin to Thales' turn toward water;[105] but this overmining reaches well beyond the current neurosciences' capacities and their abstracted vision of the brain as an ultim-ate X kind of (computational, totalizing, mechanistic) thing working alone. To embrace a radical open-endedness dedicated to emergence requires another kind of science—one more fitting to the sloshing around that allows nothing to completely dried out nor free from the embodied feeling of being absolutely soaked.

The brain of the past is nowhere and everywhere

Brains will never be understood. The brain is not the The Brain of popular culture, nor is it the other brain of scientific representation, nor that funky brain made for the museum. (So many brains!) The multiplicity of each and of phenomenal experience confuse as they overlap and brew strange fusions. We cannot settle a question of our own existence in a brain carved from the brittle steel of technicity. We might now be able to diagnose more medical conditions, cure more diseases, and provide better palliative care, but we will never see a brain nor ourselves.

We can play a social antagonist here by imagining that we will be more prosperous if adopting the mystical no-brain of the ancient Egyptians or Greeks. Tear it all down? The suggestion is starkly opposed, of course, to the neuroscientific enterprise aiming to provide "the techniques to live a responsible life" and the capacity to "manage ourselves" and find reliable means of "self-improvement."[106] Believing that neuroscience can help us "respond positively to fairness and [form a] commitment to others" and grow a more equitable society through interpretations of neurobiology relies upon enthusiasm for new imaging techniques. One must also trust the capacity of the neurosciences to overcome social meanings and power configurations when producing and reviewing brain data to hope to do such a thing.[107] But data is never "raw." Data is only ever as raw as sushi, which has been hauled to shore by fisherman with loads of gear, trimmed for the consumer, displayed on illuminated shelves, sprayed down, and packaged like jewels of yumminess. Like the raw data of the brain, sushi is never outside the temptations of the flesh.

The history of the brain concept is a history. It is a history reflective of changing times. The brain way back then is opposed to The Brain, that is the organ of explanatory Power and Might. What we see in historical lineages, then, are different bodies, some more spiritual, holistic, or connected; yet, each retain social orientations that present living flesh needing to be reconciled with its own yearnings, and often with a wider world. We cannot in early or pre-brain concepts see one organ overtaking the human. What are we able to see today? We might choose to look to this history to predict the eventual dispersal of The Brain and a return to the interconnected body at a time when we are subject to violent environments not so easily controlled.

But predicting the falling apart of power relations constructing The Brain or launching any superhero call to rethink how neuroscience is enacted to discover new ways for the brain to one day Become is not a call to abandon scientific inquiry. Rather, what feels right is a dedication to breakthroughs from wider expanses, greater levels of uncertainty about what governs us, and the development of alternative, ecological protocols. Shifting toward an open-body concept that is not neuronal nor always explanatory revisits the birds and the fish inscribing the body as in Gwion Gwion rock art. It reconsiders the mutuality of the body presented within Yin-Yang interconnectedness. It learns from da Vinci's overly rigid grid of lines and replaces an unbridled enthusiasm for current neuroscientific methods with another practice looking for embodied presences and the immediacy of ontological performances, events of Becoming, instantaneous emergences, slippages and even losses.

I crave a new neuroscience not studying the brain as singular and free nor one that thinks the body or any part as a discrete entity. Without reconsideration of what a body can mean when not having a brain, and what it can then do, nothing of liberal humanism will fall away nor will the somatic-mind split opened by Descartes undergo a final suturing. This is an abstract philosophical statement as well as a literal one. What we are likely to encounter in any technocratic, neurocentric, isolationist pursuit is another deeply carved edge drawing distinctions between human and nonhuman, one uncongenial to a future with transformations well outside the range of a body perceived as ruled by an organ.

Notes and references

1 Gert-Jan Lokhorst, "Descartes and the pineal gland," *The Stanford Encyclopedia of Philosophy* (2017). Available at: https://plato.stanford.edu/archives/win2017/entries/pineal-gland.
2 Lior Appelbaum, Gordan X. Wang, Geraldine S. Maro, Rotem Mori, Adi Tovin, Wilfredo Marin, Tohei Yokogawa, Koichi Kawakami, Stephen J. Smith, Yoav Gothilf, Emmanuel Mignot, and Phillippe Mourrain, "Sleep–wake regulation and hypocretin–melatonin interaction in zebrafish," *Proceedings of the National Academy of Sciences of the United States of America* 106, no. 51 (2009): 21942–21947.
3 Nicholas Temple, "Plotting the centre: Bramante's drawings for the new St. Peter's Basilica," in *Recto Verso: Redefining the Sketchbook*, edited by Angela Bartram, Nader El.Bizri, and Douglas Gittens (London: Routledge, 2016), 73.
4 See John Shotter, "The social construction of our 'inner' lives," *Journal of Constructivist Psychology* 10, no. 1 (1997): 7–24.
5 See Roy F. Baumeister, "The self," in *The Handbook of Social Psychology*, Vols. 1 and 2, 4th Edn, edited by D. T. Gilbert, S. T. Fiske, and G. Lindzey (New York: McGraw-Hill, 1998), 682–684.
6 Christopher M. Stratman, "Religion, art and myth-making: the halo as an aesthetic expression of ultimate reality," *JCCC Honors Journal* 2, no. 1, Article 4 (2011): online. Available at: https://scholarspace.jccc.edu/honors_journal/vol2/iss1/4.
7 Darren Curnow, "Aboriginal history rewritten again by ignorant political class," *The Conversation*, January 29, 2015, para. 6.
8 Candida Baker, "Why scientists are intrigued by the Gwion Gwion rock art of the Kimberly," *The Sydney Morning Herald*, October 20, 2016, para. 6. Available at: www.smh.com.au/lifestyle/why-scientists-are-intrigued-by-the-gwion-gwion-rock-art-of-the-kimberley-20161018-gs4zlh.html.
9 See Ian J. McNiven, "The Bradshaw debate: lessons learned from critiquing colonialist interpretations of Gwion Gwion rock art paintings of the Kimberly, Western Australia," *Australian Archeology* 72, no. 1 (2011): 35–44.

10 David Wroth, "Aboriginal rock art of the Kimberly: an overview," *Japinka Aboriginal Art*, 2017, paras 11–12. Available at: https://japingkaaboriginalart.com/articles/kimberley-rock-art-overview/.

11 McNiven, "The Bradshaw debate," 38.

12 For discussion, see: Edwin Clarke and Jerry Stannard, "Aristotle on the anatomy of the brain," *Journal of the History of Medicine and Allied Sciences* 18, no. 2 (1963): 130–148.

13 Elizabeth Sondhaus and Stanley Finger, "Aphasia and the CNS from Imhotep to Broca," *Neuropsychology* 2, no. 2 (1988): 88.

14 Ibid.

15 Charles, G. Gross, *Brain, Vision, Memory: Tales in the History of Neuroscience* (Cambridge, MA: MIT Press, 1999), 2.

16 See The Editors, "Imhotep," *Britannica*, August 24, 2018. Available at: www.britannica.com/biography/Imhotep.

17 M. P. T. Linzey, "The duplicity of Imhotep stone," *Journal of Architectural Education* 48, no. 4 (1995): 260.

18 Ibid.

19 Linzey, "The duplicity," 263.

20 Ibid.

21 George Hart, *Routledge Dictionary of Egyptian Gods and Goddesses* (London: Routledge, 2005), 133.

22 See Richard Covington, "King Tut: the Pharaoh returns!" *Smithsonian Magazine*, June, 2005. Available at: www.smithsonianmag.com/history/king-tut-the-pharaoh-returns-75720825/.

23 For more, see Staff, "Pharaoh's feminine figure explained," *Popular Science*, January 5, 2009. Available at: www.popsci.com/scitech/article/2009-01/pharaoh%E2%80%99s-feminine-figure-explained.

24 Ibid.

25 Mary Abram, "The power behind the crown: messages worn by three New Kingdom Egyptian queens," *Studia Antiqua* 5, no. 1 (2007): 5–6. Available at: https://scholarsarchive.byu.edu/studiaantiqua/vol5/iss1/4.

26 Jaromír Málek, *Egyptian Art (Art and Ideas)* (London: Phaidon Press, 1999), 266.

27 See: Gayle Gibson, "Egyptian hieroglyphs: the words of the gods," *Magazine of the Royal Ontario Museum* 45, no. 2 (2012): 24–29.

28 Calvin W. Schwabe, Joyce Adams, and Carleton T. Hodge, "Egyptian beliefs about the bull's spine: an anatomical origin for Ankh," *Anthropological Linguistics* 24, no. 4 (1982): 445–479.

29 Guenter B. Risse, "Imhotep and medicine: a reevaluation," *Western Journal of Medicine* 144, no. 5 (1986): 623.

30 Ibid., 622.

31 Ibid., 623.

32 Ibid., 622.

33 Mark Sprivak, "Neual sufficiency, reductionism, and cognitive neuropsychiatry," *Philosophy, Psychiatry, & Psychology* 18, no. 4 (2011): 339–344; Also see: Alva Noë, *Out of Our Heads: Why are You Not Your Brain and Other Lessons from the Biology of Consciousness* (New York: Hill & Wang, 2009).

34 David P. Silverman, *Ancient Egypt* (Oxford: Oxford University Press, 2003), 125.

35 See John Baines, "On the status and purposes of ancient Egyptian art," *Cambridge Archeological Journal* 4, no. 1 (1994): 67–94.

36 The origin and purpose of the image is unknown. It may be that the image was authorized by Ptolmy the reigning King of Egypt who supported medical investigation.

37 Andrew Wickens, "Five depictions of the brain," *The psychologist …* 30 (2017), online. Available at: https://thepsychologist.bps.org.uk/volume-30/april-2017/five-depictions-brain.

38 Ibid., para. 4.

39 See Tomaz Manzoni, "The cerebral ventricals, the animal spritis, and the dawn of brain localization of function," *Archives Italiennes de Biologie* 136, no. 2 (1998): 103–52.

40 Glenn Harcourt, "Andreas Vesalius and the anatomy of antique sculpture," *Representations* 17, Special Issue: The Cultural Display of the Body (1987): 37.

41 See Lokhorst, "Descartes," Sec. 1.1.

42 Wickens, "Five," para. 5.

43 Justin Skirry, "Rene Descartes: the mind-body distinction," *The Internet Encyclopedia of Philosophy*. Available at: www.iep.utm.edu/descmind/ (accessed October 15, 2018).

44 René Descartes, *The Philosophical Writings of Descartes*, Vol. 2, translated by John Cottingham (Cambridge: Cambridge University Press, 2013), Sixth Meditation, 54.

45 The idea of adding something foreign to materiality is explored by David Cecchetto, "Deconstructing affect: Posthumanism and Mark Hansen's media theory," *Theory, Culture & Society* 28, no. 5 (2011): 3–33.

46 Patricia Edmonds, "How an asexual lizard procreates alone," *National Geographic*, November issue, 2016. Available at: www.nationalgeographic.com/magazine/2016/11/basic-instincts-whiptail-lizard-asexual-reproduction/.

47 Heidi M. Apel and Rex B. Cocroft, "Plants respond to leaf vibrations caused by insect herbivore chewing," *Oecologia* 165, no. 4 (2014): 1257–1266.

48 William Uttal, *Mind and Brain: A Critical Appraisal of Cognitive Neuroscience* (Cambridge, MA: MIT Press, 2011), 3–4.

49 Ibid., 3.

50 Ibid., 4.

51 This is asserted by Bennett and colleagues. See Maxwell Bennett, Daniel Dennett, Peter Hacker, and John Searle, *Neuroscience and Philosophy: Brain, Mind, and Language* (New York: Columbia University Press, 2009), 19.

52 Uttal, *Mind*, 3.

53 "Origins of Chinese medicine," *Sacred Lotus: Chinese Medicine*. Available at: www.sacredlotus.com/go/foundations-chinese-medicine/get/origins-history-chinese-medicine (accessed October 15, 2018).

54 Today, many practicing Chinese medicine do work with neuroscientists to test herbal treatments on neurological conditions such as Alzheimer's. For an example, see: Cuiru Lin, Zhen Zhou, Qian Li, Jiakui Guo, Yulian Zhang, Jianchun Xu, Miaomiamo Long, and Di Wu, "Changes of brain activity during a functional magnetic resonance imaging stroop task study: Chinese herbal formula in Alzheimer's disease," *European Journal of Integrative Medicine* 16 (December): 46–53.

55 "Application of Ying Yang theory in Chinese medicine," *Shen-Nong.com*, paras 1–3. Available at: www.shen-nong.com/eng/principles/application1yinyang.html.

56 See: Manfred Porkert, *Theoretical Origins of Chinese Medicine* (Cambridge, MA: MIT Press, 1974).

57 Kaoru Sakatani, "Concept of mind and brain in Traditional Chinese Medicine," *Data Science Journal* 6, no. 7 (2007): S222.

58 Ibid., S223.

59 See this chapter, "brain scan 4."

60 Picture reproduced and featured also in Rolando F. Del Maestro, "Leonardo Da Vinci and the search for the soul," American Olser Society. Available at: http://cms.cws.net/content/americanosler.org/files/2015%20McGovern%20Lecture%20Booklet.pdf.

61 Charles G. Gross, "Leonardo da Vinci on the brain and eye," *The Neuroscientist* 3, no. 5 (1997): 349–350.

62 James S. Ackerman, "Leonardo da Vinci: art in science," *Daedalus* 127, no. 1 (1998): 210.

63 Ibid., 214.

64 Leonardo da Vinci, *Elements of the Science of Man*, translated by Kenneth D. Keele (London: Academic Press, 1983), 238.

65 Ackerman, "Leonardo," 209.
66 Ibid., 209–210.
67 Jonathon Pevsner, "Leonardo da Vinci, neuroscientist," *Scientific American Mind* 16, no. 1 (2005): 89.
68 Bruno Latour, "On actor–network theory: a few clarifications plus more than a few complications," *Philosophia* 25, no. 3–4 (1996): 48.
69 Ibid., 49.
70 See Donald Hoffman, *The Case against Reality, Why Evolution Hid the Truth from Our Eyes* (New York: W. W. Norton & Co., 2019).
71 Staff, "The pineal gland and the third eye chakra," *Gaia, Andye Murphy Blog*, May 6, 2016, paras 23–24. Available at: www.gaia.com/article/pineal-third-eye-chakra.
72 Katherine Hayles, *How We Became Posthuman: Virtual Bodies in Cybernetics, Literature, and Informatics* (Chicago: University of Chicago Press, 1999), 5.
73 Ibid., 43.
74 Staff, "The pineal," para. 23.
75 Davi Johnson Thornton, *Brain Culture: Neuroscience and Popular Media* (New Brunswick: Rutgers University Press, 2011), 4.
76 Ibid.
77 See Robin McKie, "No death and an enhanced life: is the future transhumanist?" *Guardian*, May 6, 2018; Sarwant Singh, "Transhumanism and the future of humanity: 7 ways the world will change by 2030," *Forbes*, November 20, 2017.
78 Kimerer L. LaMothe, *Nietzsche's Dancers: Isadora Duncan, Martha Graham, and the Revaluation of Christian Values* (New York: Palgrave Macmillan, 2006), 219.
79 See: Dylan Thomas, "Do not go gentle into that good night," *Poem Hunter*. Available at: www.poemhunter.com/poem/do-not-go-gentle-into-that-good-night/.
80 Tim Sharp, "Alpha Centauri: nearest star system to the sun," *Space*, January 18, 2018. Available at: www.space.com/18090-alpha-centauri-nearest-star-system.html. For discussions of dark matter, see: "Dark energy, dark matter," *NASA: Share the Science*. Available at: https://science.nasa.gov/astrophysics/focus-areas/what-is-dark-energy.
81 Benedictus de Spinoza, *Ethics, Part III* (Project Gutenberg, 1677). Available at: www.gutenberg.org/ebooks/948 (accessed September 20, 2018).
82 Staff, "The pineal," para. 24.
83 Gross, *Brain*, 10.
84 Reginald E. Allen, *Greek Philosophy: Thales to Aristotle* (New York: The Free Press, 1991), 2–3.
85 Ibid.
86 Ibid., 3.
87 Graham Harman, *The Quadruple Object* (New York: Zero Books, 2011), 6–11.
88 Allen, *Greek*, 3. Note: the passage appears in Allen but derives originally from Aristotle's comment in *Physics* 204b 24, drawn from John Burnet's translation.
89 Ibid.
90 Harman, *The Quadruple*, 6–11.
91 See "Walking contradiction, Green Day," *Genius*. Available at: https://genius.com/Green-day-walking-contradiction-lyrics.
92 Gross, *Brain*, 10.
93 William Schultz, "Neuroessentialism: theoretical and clinical considerations," *Journal of Humanistic Psychology* 58, no. 6 (2015): 608.
94 Peter Reiner, "The rise of neuroessentialism," in *The Oxford Handbook of Neuroethics*, edited by Judy Illes and Barbara J. Sahakian (Oxford: Oxford University Press, 2011), 161.
95 See Harman, *The Quadruple*, 6–11.
96 Nikolas Rose, "Reading the human brain: how the mind became legible," *Body & Society* 22, no. 2 (2015): 158.

97 Nikolas Rose, "Human sciences in a biological age," *Theory, Culture & Society* 30, no. 1 (2013): 23.
98 See Kenneth Burke, "Terministic screens," in *Language as Symbolic Action* (Cambridge: Cambridge University Press, 1966), 45.
99 Jeff Coulter, "Twenty-five theses against cognitivism," *Theory, Culture & Society* 25, no. 2 (2008): 20.
100 Ibid.
101 Ibid., 21.
102 The term "neuro-chemical self" derives from Rose. See Nikolas Rose, "Neurochemical selves," *Society* 41, no. 1 (2003): 46–59.
103 Clifford van Ommen, and Vasi van Deventer, "Negotiating neuroscience: Le Doux's 'dramatic ensemble,'" *Theory & Psychology* 26, no. 5 (2016): 573.
104 This section is drawn from a discussion in my article on this topic. See David R. Gruber, "There is no brain: Rethinking neuroscience through a nomadic ontology," *Body & Society* 25, no. 2 (2019): 56–87.
105 Rose, "Reading," 158.
106 Rose, "Human," 23.
107 Ibid.

2
NEUROSENSUALITY AND BRAIN ART

Neurosensuality crafts an intimacy with neuroscience. The concept references an attachment or particular pathos around the practices and outputs of the contemporary neurosciences. It welcomes the brain scan into the house and opens the heart to penetration. Neurosensuality resembles what Racine and colleagues call "neuro-essentialism," or the need to squeal, "I *am* just like that!" when hearing a new neuro-description.[1] But neurosensuality focuses less on neuroscience outlining our personhood; it focuses more on producing guidance and comfort through delineating the human-as-a-brain. In neurosensuality, neuroscience is not cold and clinical; it is a warming blanket in the face of psychological distress, social disturbance, and the instability of the subject.

In brain art, neurosensuality appears at the intersection of neuroscience and everyday environments, promising deeply moving revelations about the ways that our brains relate to family troubles and personal calamities. Creatively, neurosensuality functions as a thematic. In the three artworks explored in this chapter, it comes into view when artists stage the material Things of neuroscience—the scans, scalpels, and electrodes—in relation to the places where feelings about everyday life are tangible, i.e. in memory boxes, blankets, and souvenirs. These sentimental items then speak and say more about us or our condition when viewed and understood through neurobiological discourses. In reverse, those things set neuroscience into an ethical relationship with family life and personal well-being. They ask questions of the neurosciences: How is it caring? Who does it affect? Can it aim to improve both health and well-being?

Scan 5: the brain keeps the body above the break

Philosopher Walter Stace notes two kinds of mystical experiences common to "all cultures, religions, periods, and social conditions," which he terms the "extrovertive and introvertive."[2] The extrovertive pushes out and experimentalizes to "apprehend the One or the Oneness of all in or through the multiplicity of the world," whereas the introvertive focuses on the inner manipulation of "concepts, thoughts, sense perception, and sensuous images."[3] The dualism, regardless of its simplicity, reveals a broad-ranging tendency: to imagine the world (or layers of worlds) divided from us. To look inward or to reach outward is a bifurcated image; a human must bridge a gap while living in isolation. Yet isolation is a strange seduction; for even when we reimagine ourselves within the world and highlight our visceral integrations and vibrating comportments to varied and moody environments, we dream nevertheless that we will not be lost in a muddle. That is, we intend, come hell or high water, to Be and Be defined. We must distinguish ourselves. We aim to determine how, exactly, we are a Thing, even if "extended" or systematized or networked or in some process of changing, slowly, if inevitably.[4] We want to know how we can fit ourselves into a much bigger picture.

Particularity triumphs repeatedly over open-endedness. Why shouldn't it? The tangible instance—shaping clay, growing vegetables—offers the reassurance that we can do something, enact some agency of change. And perhaps we can, sometimes. But the broader world nags at the suggestion ever being a generalizable principle.

The dynamic and diffuse technological life intensifies the feeling of instability in Being. Instantaneous regeneration of digital media and the confounding nature of technological complexity showcase our mental limits and the impenetrableness we sense in the world. The "always on" disrupts our sleep; the specially tailored feeds bother our securities; the constant fakes and questioning of information sources expose the depth of our ignorance. Who knows what information comes from where? Did an algorithm write that message? Are those images real? Is a star crashing toward earth? The sand bar shifts.

If we are going to jump into the water, we need assurance that we're swimming somewhere, somewhere with an island of coconuts and sunscreen. We're going to need loads of sunscreen. It is 90 degrees in Copenhagen today.

Nobody feels very comfortable alone out at sea. Black water swirls. We struggle to keep the body above the break. What Sartre calls "the quietism of despair" grows louder.[5] Our existential condition cannot be confined to the staid halls of contemplative philosophy. The "moment of nervousness" cannot be ignored amid the moment to moment report. The anguish of living in new media is not the feeling of losing "an *a priori* existence ascribed" following after the (now dull and overplayed) philosophical rejections of grand religious authority; rather, it is no longer having one minute's worth of felt stabilization in our social and political environments.[6] We must now push beyond the demure academic concerns about fragmentation and loss of ground for interpretation addressed by postmodernism's socialites.

Non-linearity, unreliable narrators, and subversive plots constitute the end of last century's literary responses to a philosophical rejection of grand narratives and dispersal of formalistic approaches deemed too reliant on universal structures. Today, iterative fiction, new media literatures of surprise, operating algorithmically beyond the control and understanding of the author, are more fitting responses to an exceedingly complex and raucous technological milieu. Literary works by John Cayley, Daniel Howe, and Jhave Johnson offer interactive, performative experiences whose content is dispersed across data sets, suggesting that what burrows into our heads are not single speeches or paradigmatic images, but a blushing rush of various media, constantly, relentlessly changing.[7] To see those works once is to see them algorithmically generating a unique combinatory moment; no one will ever see them the same way again. Mirroring wild social configurations and dynamic information flows, they are visibly dismantled moment to moment, whether the viewer likes it or not. So they push us to now pay attention to material tumult and consider social fragmentation not only as old news[8] but as a modest and even cute observation.

The social and political task is no longer to discover secret all-encompassing ideological formations that dominate the masses in equal respect—did ideologies ever function so flatly and with that kind of domineering unity anyway? The task is to keep up with the formation processes, to discover the salient agents, human or nonhuman, and try to swim. Yet we face the inability to take responsibility for the iterative, re-combinatory composition when the responsibility remains yet ours.

Sartre argues that a recognition that each person is not "a particular example of a universal conception" comes with "the sense of complete and profound responsibility" for ourselves and humanity; although he overlooks just how much religious universalizing also seeks to impose its own depth of/for human responsibility, he points to a problem, namely, that humans feel "profound responsibility" but also struggle to define it in a manner fitting to humanity's divergences and multiplicity.[9] Determining a path and enacting a plan, especially at scale, sits idly on the spinning wheel of the thinking computer, subject to "refresh." Dispersal, lack of access to information systems, to processes for storage and retrieval, and to numerous material re-integrations complicate any responsibility; unseen actors can be blotted out at the turn of a dark satellite. Algorithms revise. Networks reform. Hackers and security teams work behind the scenes. Sea cables and phone towers relay innumerable signals. The air vibrates. Governments and corporations collect, spy, and adapt. Relentless change and emergence fashion the present as already historical. History is now a data set whose shape is written by robots.

No one can escape emotional disruption: incessant alternative realities, fake news, sudden economic breakdowns, climatic oppressions, eruptions of eruptions. "God, it really is hot today." If "no man is an island" seemed appropriate once, then "no island is an island" fits today. Every wondrous and seemingly meaningful proposal, whether intervening at the ontological or epistemological level, faces an order of magnitude of uncertainty, right inside of daily life, greater, no, yes, than ever before.

Anxiety, a recent *Independent* news article declared, is "society's prevailing condition."[10] The question is not whether we can prop up an articulation of what we are, as a people, a community, but whether we can feel secure in any articulation. Perhaps we see something in the unpredictability of the weather, the rapidity of news cycle, and the constant surprise of the data stream. We see a performance of existence. We cannot help but remember what we have been taught: existence comes before essence. Google God, and the conditions of existence shuffle around; so will we.

Here is where the brain and brain art have a voice. They shout contradictory things at each other and then sometimes switch sides in a debate about the future.

We, ourselves, a brain scan image announces, need not be strange simply for being complex. We need not live with ontological anguish and anxiety. We can find real, workable, reliable solutions from the solid ground of our own neurobiology. Brain art repeats the equation, but the best of it also taunts the summation.

Putting a brain in a museum foregrounds a recurring function for the museum space: the personal illumination. We hope that smartly framed artefacts will tell us more about ourselves. The big white museum space becomes, in this sense, a giant brain scanner, and especially when the brain is introduced. Attending a brain art exhibit, the "us" under scrutiny is not the vague and all-encompassing "social" that critics banter about in the lobby when viewing paintings; in this case, the "us" is framed as the "real" us, the scientific materiality making us and showing us to have a common, a universal.

Thinking in juxtaposition to a painter like Vasily Kandinsky provides a crisp contrast: he paints geometric forms with an aesthetic pleasure, which test the limits of collective social sensibilities about art. His works revolve around abstract questions, and they do not require that audiences look for the specific psychological tendencies of the artist. His images do not treat the artist as an individual; they certainly do not ask viewers to transpose the art onto the Self.[11] But brain scans staged alongside of personal medical narratives or sculptures of brains made with objects taken from an artist's home prove quite different. They are uninterested in formalistic play, even if the artists, like Kandinsky, remained "convinced that content could be communicated by line, color, movement and direction."[12] Brain art, again like Kandinsky, also tests our collective sensibilities about art, but it authorizes us to reconfigure our subjectivities—neuroscience seen in personal domains suggests that we can (or should) apply this and be put at ease about technological capacities illuminating our secret loves and longings. Only if we are charmed or enamored by this technical drive toward confessionalism—or is it a drive toward greater institutionalized surveillance or maybe toward a deconstruction of the Self?—will the exhibit be at all acceptable. Neurosensuality in brain art, we might then say, composes a scientific shape for intimacy.

The three artworks explored in this chapter compel us to (re)think how neuroscience influences how we view ourselves and how we imagine everyday life. Most artworks suggest that neuroscience is exceedingly useful—for remembrance, for expressions of mourning, and for reflections on important emotional events. But this is but a simplistic starting point. They also—some anyway—tacitly endorse neuro-centric and limiting perspectives, even while all of the artists, by and large, approach questions of personhood and identity seriously, meaning in unresolved ways, and treat people with the necessary respect. The artists understand that we need means to process an emotional life. They see how we cannot easily escape our medical, institutional structures and must, rather, work with/in them. They recognize that biomedical tools and discourses shape how we learn of neurological diseases and make a difference to how we deal with problems like devastating memory loss. Accordingly, the works discussed in this chapter encourage a neuroscience that is more sensitive to living with a body that must be measured, scanned, datafied, and medicalized to live.

Brain scans can change people's feelings about themselves and redirect the course of their lives. Brain art recognizes this. The artworks in this chapter seem to argue that neuroscience should not ignore neuroscience's effects on the emotional life. If these artworks are at all judged for placating neuro-essentialisms or offering uncritical assessments of neuroscience's power, then they achieve abolition of sin in honest endeavors and emotional attentiveness. Technicians are reframed as sages or saints; patients become sojourners; brain researchers hold the keys to heaven or hell. Through intimate imagery, we discover an ethical dimension to studying the brain's systems, cataloguing conditions, and diagnosing individuals.

The works, I will also briefly note, are instructive to the act of criticism. They help critics to better account for sensual understandings and see the world through the body. So I try to take the lesson to heart—to think from my own body. I test my emotional comfort zones and consider how the works propose modes of living with a technical apparatus that can peer into brains. If we decide not to be content to view brain art as a simplistic celebration of hidden neuronal "truths" sitting right beneath the surface, then we must dive into the psychological depths; we need to try to swim the heavy ocean of human finitude washing ashore our own mortal flesh. Looking at an image of the brain offers the chance to wade into our own little mucky pond of cerebral comprehensions. We hope to emerge more compassionate after dipping into neuronal life.

Knitting a brain: *Warm Glow*

Red, blue, purple, yellow, and orange lines run horizontally across the visual field of *Warm Glow*, a 5X6 wool rug knitted by Marjorie Taylor.[13] In the foreground, the same bright colors compose the bulbous shapes of brain regions pictured as a horizontal brain slice. The pattern is familiar, dictated by the functional magnetic resonance imaging (fMRI) machine as it scans a brain front to back. But this is no run-of-the-mill brain. Taylor weaves the warm glow of her husband Bill's brain.

As a rug, spread out on the floor, we can walk right over Bill—or is it his brain? The idea of stepping on his frontal cortex feels uncomfortable. Neuroscience, we remember, intends to visualize Bill, or rather, to show us his brain in a picture; stepping on the rug feels like trampling all over Bill.

Touching the rug is a modality shift that forces an emotional response. Bill seems suddenly closer. Are we intruding? Does the brain scan show too much? Is Bill sick? We wonder if we need to talk to Bill about his brain were he sitting in a lounger in the living room. What a conversation piece! "Hey, Bill, how's the old brain? Got a tumor? Looks all right to me. I'm no doctor, but hey, that rug sure looks snazzy ..." That would be insensitive, we remind ourselves. But it seems unlikely that his wife would put his entire brain scan out on display if he really confronted any serious danger. Whatever the case, we feel cognitive dissonance stomping all over Bill—or is it his only brain? We're still confused.

We would probably never stretch out on this rug on the living room floor with a cool glass of Chardonnay, bopping our heads to a Duke Ellington tune. What if we spilled the wine?! Good news, Bill, it was a white, not that awful Merlot! A little soda water should do the trick!

The rug forces a reorientation to the brain scan, or our idea of it. The visually pleasing symmetrical shapes add to the effect, compelling us to look differently, closer to the way that we view Persian and Oriental rugs. We see the brain now as a design, as a blend of colors, not a unified neuroscientific object. In like manner, the fMRI scan—or is it Bill's brain?—can no longer be a tool for medical diagnosis; it transforms into a household item meant to comfort us. We can feel good about it. We try to, at least.

The intimacy of the rug—of rugs themselves, but especially of a rug taking a brain as its subject matter—suggests that there is something truly sensual about the fMRI process. We wonder if neuroscientists feel the same way about scanning brains as artists do about making rugs. Do they give it such tender care? Do they spend massive amounts of time scrutinizing over the colors and textures? We guess that they don't.

But the rug insinuates that the scanning process is itself an experience worth honoring. Bill was, after all, inside of the scanner. All of the secrets of his brain, what lay at the foundations of complex mentalizations, his ideas about family and friends and his habit of licking the back of the spoon, might well be materially notable. We can imagine a neuroscientist gossiping in the back of the laboratory: "Hey, John, check it out. The orbital frontal cortex on that guy doesn't light

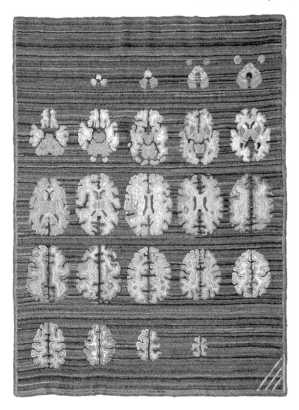

FIGURE 6   *Warm Glow*, Fabric MRI: Bill's Brain, 2009, Marjorie Taylor.

when he fills out the form. And the thalamus is inert. Yikes!" Why not give this neuroscientific process, this intrusion, the intimacy it deserves? Why not craft it with the labor of hands, over many arduous days, so that it spreads out slowly over time to cover the hardwood floors?

This touching re-mediation requests a new emotion-attuned vision. *Warm Glow* offers another way to see Bill. But it is not only or even most forcefully his brain that we see; what we see are ties to his family and the troubles that he may face as a result of this scan. Indeed, if brain scans are visually "seductive,"[14] then *Warm Glow* asks us to be drawn in by the emotional curiosity of the circuitry, to displace the specular fascination of flesh with the living, affective embodiment of it. Touch the rug's soft fabric, Taylor seems to say, and be seduced by materialization—but do not fetishize scientific materiality; see the material of the home and bask in the glow of the heart.

In a literal and figurative sense, the rug places neuroscience underfoot. Lowered from its technical throne of celebrity rock stardom, the scan becomes a resting place for the dog or cat. Yet this is deceptive. The tension of the rug relegated to the cold floor stirs within the hot necessity of neuroscience for Bill.

The power of the rug is wondering if, in fact, Bill suffers from a debilitating neurological problem and how much this scan has actually changed his life, where it will all lead, and whether the artist mourns in this way. *Warm Glow* leaves this question of suffering unanswered. All we have is the inculcation of the scan now made to be in a place where everyday life must go on.

By inviting us to warm our feet on the glow of Bill's brain scan, we start to feel what neuroscience *does for people* once outside of the lab. Once picturing the scan inside of the home of those who stare at their own brain scans and try to make sense of them, we think how Bill must live with this artefact. We realize that he has to deal with whatever emotional or physical consequences it represents. Why not cuddle up to it?

The personal touch glows most warmly in *Warm Glow*. The viewer is seduced by the time and tenderness needed to knit the detail of the rug—the neuronal complexity of one's own husband's brain—and to knit so lovingly in those soft colors. The rug, we imagine, more completely unites Marjorie and Bill. Bill is loved, and we cannot help but be a witness. Through knitting or through immense amounts of time spent caring for Bill, Marjorie gets to know every corner of Bill's brain. Every memory tucked into every fold. Here Marjorie—as the artist or as Bill's wife?—quilts her own view of Bill—or is it his brain? Unavoidably, she quilts herself into the representation; the work is identifiable as her own. But more sensually, she may well embed her own memories of them together into the meaning of the rug, irrevocably stitching their time into the fabric of the home. By lacing together wool fabric strips, the rug becomes a visible symbol of a life built together. *Warm Glow* manifests their collective materiality and destiny.

Seen from a different angle, *Warm Glow* plods right along with traditional femininity where quilting and knitting are domestic, utilitarian practices, and the wife's task is to care for the home and the husband. Taylor, we are tempted to think, spends day and night considering Bill's thoughts, caring for his needs. However, *Warm Glow* cannot be so traditional. The artwork is not so simple as to be about a glorious husband and a subservient wife. The work, as a craft and an unexpected form of brain art, carries also a kind of protest against traditionalism.

Pentney reminds us that crafts like knitting can be a "feminist act" strategically used to question old hierarchical structures.[15] Knitting has been "denigrated" and staged as "traditionally woman-centered activity," which, consequently, opens possibilities for reclamation through subversive meaning associations. Recently, knitting has been used to express a fusion of "fun and politics" while questioning gendered practices and assumptions in efforts to revise traditional subject positions.[16] The social and political force of knitting, in brief, works from the expectation of its alignment with a traditional femininity, functioning as a mode of resistance when breaking out of its traditional forms and uses.

Choosing where and what to knit does much of the political work. Beryl Tsang, as one example, knits breast prosthetics available for women with breast cancer[17] to "provide a way for women to reconnect with their bodies through craft" but also to call attention to the odd, sometimes funny or unconventional, choices that women must make when needing to choose consumer products tailored for/ to their bodies.[18] Making fuzzy organs or knitted breasts requires interpersonal reflection on medical practices but also interrogates women's choices when forced to re-imagine and re-engage their own bodies through prostheses, even as they need to heal from potentially devastating disease. Making the works available in online shopping venues and having women wear them as actual prostheses focuses attention on the consumerism driving the industry as well as raises critical questions about care in modes of production around women's health.[19]

Although it is admittedly difficult to see *Warm Glow* as an explicitly political project to generate "the spirit of feminist goals of empowerment,"[20] the work, nevertheless, functions politically and subversively. The historical domesticity and denigration of quilting makes the act of quilting a brain a challenge to the sciences of "Man" dominated by men in white lab coats. *Warm Glow*, in particular, positions quilting as a practice able, just the same as neuroscience, to generate serious reflection on material structures that contribute to important life experiences.

To go yet further: *Warm Glow* undermines neuroscience's claim over Bill's materiality. The rug presents an alternative to what Bill means. *Warm Glow* suggests that the interior of Bill's brain, like the home, cannot be captured in any single form of material engagement. Bill's brain is multiple. And Marjorie or Bill himself may well be a better judge of Bill and what he needs. Fabric art, in this way, protests ontological reduction and certitude. The rug suggests that rote medicalizations of abnormality might not be a good fit for Marjorie and Bill; perhaps they would rather stay at home cuddling by the fire.

Quilting an fMRI brain scan can be seen as a revisionary act, one challenging reigning orders of epistemological politics. If knitting is traditionally an unnoticed and unimportant activity receiving little cultural celebration, then neuroscience is its sizeable opposite. The juxtaposition creates a realignment and declaration of validation. It also introduces uncertainty about who knows what and how anyone can come to know a brain—or is it Bill? We start to wonder if neuroscientists can see Bill or whether they intentionally conflate him with his brain to increase the field's epistemological power. The big seeing-eye of neuroscience—an institutional spectator standing at some distance from Bill's everyday life—might as well warm the floor, the artist seems to say. The brain scan is, in a material and symbolic way, delegated to the realms of knitting as it proves quite deficient to the task that it gives itself.

An additional epistemological challenge implied by the work is rethinking the educational value of traditionally domestic material explorations. Practices like knitting and quilting have been ordered as "old-fashioned" and "solitary pursuits," not as a valued means for learning through arts in the same way as, say, painting has been.[21] For hundreds of years, painting was used for the instruction of Biblical texts or to teach morality tales.[22] As one might imagine, it is no exaggeration to say that quilting stands, then, as something of a stark contrast (educationally, with respect to supposed value) to material investigations of the brain through scanning technologies. As a result, quilting an fMRI brain scan immediately asks audiences to look again and to see quilting as another way to learn—as a legitimate way to investigate bodies. The work encourages a lesson about what to see when looking at/through different materials and how new means of access shape the meanings that can be applied to bodies.

Together, Bill's brain and a scanning technology compose a form of knowledge, but considering how we must think about Bill's brain anew when quilting it, we might entertain the notion that material structures integrate ontologically. That is to say, neuroscience, quite literally, makes Bill's brain into a certain kind, just as Marjorie does. But the material practices of quilting, despite following the aesthetic orders of neuro-imaging, hold the advantage of time; arguably, then, her mode of engagement expands reflection. As a means of learning about a brain, quilting's long struggle toward completion—with visits to fabric shops, with long strands of yarn cut to size—compel close looking over longer periods of time and make associations that push us to ruminate on his life as a person while we map out the structures of his brain. The work here asks if neuroscientists ever look long enough or if the neurosurgeon gave Bill enough consideration.

Although the rug can be understood as an affront to the power of neuroscience to be the funnel point for knowing more about ourselves, to Taylor, as quilter and yarn artist, what may well matter most is the impact of the aesthetic of Bill's brain, not really what any neuroscientist sees. The story is an old one. What the artist cares about is the art, or the aesthetic pursuit. As Taylor puts it, "I couldn't help but look at them with the eye of a quilter ... I thought that the folds of the cerebral cortex would be great in velvet."[23] That is to say, when seeing the brain scan, the quilter or the artist or the wife—whichever should be foregrounded here—does not see Bill at all. For her, the scan does not need to be essentialized, i.e. to be "Bill." The scan could, rather, be a sensual pursuit for its own sake and a colorful expression. And the scan does, in fact, appear immediately aesthetic. It strikes the viewer as a thing whose starting point is a pleasing affect or the embodied experience of fabric.

The aesthetic interest calls attention to other potentials in *Warm Glow*. Its textures and threads highlight the selections that go into a brain scan. The neuroscientific scan was, after all, made in machines and woven in algorithmic digitality then forged through image contrasts and sharpening tools and printed in bright high-resolution. Whether the red blobs on fMRI scans are inherently seductive or are by now mundane visual objects, *Warm Glow* reminds the viewer that they are con-structions with designers and that they can be beautiful. We see once more (or for the first time) the choices and how the images are framed, enhanced, tailored for a specific purpose. In fact, because the processes used to compose the rug cannot go unrecognized in the creative artefact—its every weave and fold are inevitably open to investigation—brain scans, in turn, come a little more to surface. The choices are not so evident in the computational and clinical manifestations of a scan; there, the brain scan hides from us. *Warm Glow* reminds us of this fact.

Ultimately, *Warm Glow* punctuates the plurality of realities that can come into view when materiality is engaged differently.[24] The glow, so to speak, of an fMRI brain scan is not singular, just as the rug is not. In meshing purple, red, and yellow yarn, Taylor constructs numerous glows: the warmth of Bill's eyes, the household fireplace, the red blobs on the scan, the light emanating from an image reader in a doctor's office, not to mention a new relation to Bill. The outcome is an artwork suggesting that what makes a brain "warm"—a word attending to visual sensation as much as tactility and emotional character—depends on the material means and relations drawn to it. The hot spots on an fMRI scan light up the brain in one way, but the quilt provides another kind of glow. In the juxtaposition, we comprehend different sides of the brain—or is it Bill? It could be us that glows.

The artwork is multiple. A brain scan rug can warm us up to the idea that we can know more about our personal lives with a scan. Or, the work might ask whether neuroscience is the best place to learn about Bill and assess his personality. Or, it might compel us to think about the discourses of neuroscience as happening in the home. Or, it might show us how memories of our medical experiences are materially woven in brains, which are not only represented by neuroscience practices and then shown back to us with clinical efficiency, but ever recomposed through lived engagements. *Warm Glow* is neurosensuality.

Probing a brain: *A-Me: Augmented Memories*

Nobody likes a brain surgery. Most of us would go to great lengths to avoid it. A metallic probe piercing the skull, entering the soft, wet tissue of the brain, and poking around a bit, is cringeworthy. We pray that we never have to lie under the bright fluorescents inside of the sickly hospital room while a surgeon fires up the circular saw, then jabs a scalpel into the parietal lobe.

Surgical stations with their cleaning fluid smells and unnerving blandness are not ideal settings for picnic table stories about mom and pop falling in love. These are not places where we drum up sentimentality about the good old days. In fact, the very question of whether we can be in the mood to ruminate on our lives and share memories with others is an environmental one. We require the right place and time. A few glasses of wine, perhaps. A glorious sunset. A friendly face. We're sensitive creatures. Asking audiences to do this heavy emotional lifting while also performing a brain surgery would be an odd thing indeed. But that is exactly what Jordi Puig, Annamaria Carusi, Alvaro Cassinelli, Philippe Pinel, and Aud Sissel Heol do with their collaborative brain artwork, *A-Me: Augmented Memories*.[25]

The trick of the work is in making audiences feel comfortable. Attention to atmospheres is required. The artists, thus, restage the surgical apparatus in the comfort of a darkened museum space, among the enclosed intimacy of black curtains. Supported by calming music and following a nice glass of wine in the museum lobby, they ask audiences to interact with surgical tools in their own time, with some privacy. The activity is framed as discovering other people's cherished memories.

Audience participants, first, pick up a wireless probe and set it on a dummy's head; crucially, they do not insert the probe into the head or jam it through a skull. So the work is, in this detail, already distanced from the actual experience of the surgical bay. Even so, an audience participant needs to imagine conducting surgery and must simultaneously watch a screen with a VR display showing an fMRI scan of a brain, staged as a real-life view of a patient being probed in real time. As the probe moves over the many regions of the brain, the display changes accordingly. Viewers encounter red dots that have been placed in the brain by past participants. The dots can be clicked to open sound files containing memories. The user listens. A story of grandpa's work clothes. A snippet about falling in love. Participants, thusly enamored, are encouraged to find a special, secret place in the brain to leave their own memory.[26]

The artists describe *A-Me* as a project exploring "the ambiguity between the possibility of accurately locating places in the brain, and the uncertainty of defining a *place* in the world (or the brain) for a mnemonic experience."[27] Thus, the work highlights a debate in the neurosciences about the delimitation of memory to the brain and plays with a tension about the mind–brain relation. Namely, the project treats memories as "location-based" even while recognizing in the act being performed that "remembering is an active, exploratory process."[28] Although memories might require the firing of neurons, some memories also need external stimuli to be experienced. Asking participants to inject memories into a brain explores creatively how environments alter or bring about memories; the project experiments with what participants remember in this odd space and what they choose to add to the collective "memory palace" that *A-Me* becomes. Finding memories or placing them in the brain, on one hand, advocates the mapping practices of the cognitive neurosciences; yet, on the other hand, the artists initiate an ironical performance by asking audiences to use the exhibit to conjure memories, compelling interrogation of the idea that locating memories in flesh alone is possible.

Positioning participants as medical practitioners, surgeons, or neuroscientists, *A-Me* also broaches the uncomfortable psychological compulsion to probe other minds. To see past the curtain. To live inside another. To access the most intimate. Being a bodily transgression, a literal probing of the brain with a metallic tool, the artwork raises ethical questions about a broader human desire that may be rooted—as I note later in this book—in a craving to expose one's inner Self, that is, to be fully seen, which can sometimes manifest as the need or compulsion to see others completely naked.[29]

Tellingly, there is no indication in *A-Me* that the brain on the VR screen has given its consent. No clear storyline for the brain has been laid out before the participant. Outside of a dummy head disconnected from its own body, there is no identifiable person by which to connect to this brain. The memories are, thus, sitting idly inside of meat, extracted from an emotionally vibrant life, say, from a lovely suburban town in Northern California where the neighbors get together once a month for dinner and share details about their children and travel adventures. No, this specular human head, this flesh, in being so disconnected, is open to abuse. Extracted from ethical limitations, the brain-body is completely under the participant's scalpel. We have to embrace an optical eroticism as well as a murderous opportunity. No social sphere with community norms is inscribed as governing here. The boundaries are wide open. The tools have been put right into the visitor's hands, whomever those visitors may be. The power of erasure. Participants are encouraged to alter the mind forever—to impose their own will, inject their own memories. *A-Me* asks for a kind of mental colonization.

The absence of any obvious personhood for the patient within the surgical museum experience seems inherently a critique of neuroscience in practice. The effect is strengthened by the technical apparatus. In working through the real equipment of neurosurgical units, an ethical interrogation is forged regarding the use and development of these tools, which are presumably made to erase or modify brain tissue. We begin to wonder how much leeway the surgeons have been given, if they have authorization, and how much, or how they have obtained it. We wonder if the surgeons know what, exactly, they erase. Have they such comprehensive knowledge? Have they considered the lifelong effects on the person? What if they slip up?

The immediacy of the VR brain does not show the viewer any of the down-stream effects of brain alterations, but it, nevertheless, allows the viewer to aimlessly wander around the brain to look for glowing red dots. In this way, the technical visuality of *A-Me* helps the participants to understand the surprising way that "Brain atlases are being used" within a methodology where participants are asked to per-form concrete tasks in order to stimulate neurons absent any broader theory about brain functionality; this is a neuroscientific "black-box approach" to cracking underlying processes. We might then ask what it can mean to adopt the screen's limitations—to see in such vague terms—without an underlying theory. What the hell are we doing? In poking a red dot or adding another, we have no clue as to what kind of brain we make and what else might change forever.

The idea of injecting new and foreign memories stirs un-comfortability. At least we think it should, or hope it will. (*What if that were us laying on that operating table? Good thing we are the surgeons. Phew!*) In probing the brain and adding mem-ories, the inculcation of our own life experience, so imposed, alters the patient for good or for ill. We might think about all of those people who will follow now after us, after we finish dissecting and playing God, all those who will hear our memories. Will they be wowed, or will ours be mundane, or worst, will we be lost in muddy combination with the others swirling there? Or will the memories inherently change when materialized this new way? Perhaps we never considered that cutting them into an audio file would make them foreign. These answers we cannot know except from the perspective of those following after us—or more pertinently from the perspective of the person (or dummy non-person) whose brain we have just invaded.

Thinking about the value of our memories compels us to eventually look down; despite the dummy head, we imagine that we see a person on the table subject to our blade. We feel for them. Maybe. Or, we feel that we are somehow making this person better. Can we think so highly of own memory experiences? It is possible— isn't it?—that my memories, interjected, do not mesh nicely, but corrupt.

As might already be evident, *A-Me* is not only a project about neuroscience but about self-interrogation. The work tests our limits. It probes our own gull as well as our arrogance. It asks whether we can see ourselves and/or neuroscience, in some way, enacting a Will to Power. As a "psychological hypothesis," the Nietzschean Will to Power presents for consideration Beings "fundamentally composed of centers of power exerting force against one another."[30] The Modern anxiety about the uproar of the classes, the technological and economic fragility after the global fall of monarchies, which the Nietzschean concept captures, is herein extended out to neuroscience. In accordance with the Nietzschean reading, *A-Me* suggests that our contemporary high-tech digital means do not escape first-order impulses. The human dance is the same. Dare we so smugly pronounce ourselves as far beyond the old impulses of Modernity and living now in the future?

In the cozy corner environment of *A-Me*, our happy, cherished life memories are dislocated from their bed of comfort and called forth as a means to exercise our own Will to Power. Or, if one prefers to think closer-to-the-ground, then *A-Me* might be said to ask us to consider our lives as a pragmatic solution to the problems of another, as a biopower.[31] *A-Me* can be interpreted as strategic, reasoned, governed, legitimated tyranny over another body, an "exercise of sovereign power" and "the power over death."[32] Nietzsche's organism might, in expressing its Will, rage against the relations of things and impose itself, desiring to exploit the limitations of others, but Foucault's biopower finds the necessary means.

However, no one believes in pure evil anymore. Bad intentions are not presupposed. And any nefarious mustache-twirling characters with dark psychological compulsions to eat brains seem unlikely. Thus, we suppose that *A-Me* is rightly framed around the hype and excitement of the advances of the brain sciences, and the underlying character of a Will to Power being played out as a biopower can lose some of its terrible bully tonality. Enacted as a contemporary, interdisciplinary hybrid project where neuroscience is integrated into art practice, the force of opposition stirring within a strange situation where the museum-goer exerts control over an imagined real-life brain of another person dissipates a little. The threat fades away. Some of the underlying compulsion to expose one's own mind while gazing at the helpless Other might also drift off, unnoticed. Sharing a night out to the museum with friends, sipping a house Merlot, chomping a prosciutto-wrapped olive takes the edge off. But should the edge be sharpened?

Some of the aesthetic choices can make a discernible difference to how we feel about the work. The cartoonish dummy head appears without any human distinction. No eyes or eyebrows, void of a facial expression. Similarly, the VR display offers a flattened 2D representation. These details distance any nefarious compulsions; the act becomes emotionally flattened. To truly realize the nastiness that can be our own impulses, we probably need much more dimensionality. We need to confront the Other face-to-face to best comprehend the force of our impositions. But standing over a dummy head, staring at a VR screen, no such immediacy or conflict occurs. We can proceed without delay and with a self-assured confidence in the direction that we, ourselves, want to take this brain.

As a reflection on neuroscience and what it does, *A-Me* does not think neuroscience kind. As a science that sees, trains, and practices in specific ways with admittedly crude tools, often absent a coherent overarching theory, as the artists duly point out, *A-Me* positions ignorance and opposition at the heart of the endeavor to learn about brains. Something wildly undisciplined, more raucously practiced is forwarded. But likewise, something more personal, much closer to the everyday life of the person that eats too much, runs sluggishly, and sweats with bouts of erratic nervousness is implied as being needed to assure ourselves of ethical relations amid actions gaged in microscopic percentages through computer algorithms and metallic probes.

A-Me cannot be solely characterized as a performance of a Will to Power. The work, more immediately, foregrounds inner sentimentality about the museum-goer's own life and the memories held dear and worth sharing. In that sense, it is directed inward and toward the meaningfulness of having memories in the first place. Ruminating on one's treasured moments underscores the devastating impacts that brain diseases have and the inherent good of surgical interventions aiming to salvage whatever is left. *A-Me*, likewise, fosters self-wonder about a human body where alterations to specific regions make a difference to how we think of ourselves and what we can remember. Here is where we feel the most productive tension: an unresolvable opposition between imposing our own Will and rescuing that of another. Of being rescuer and being destroyer.

Once imagined as if having the unsurveilled power to reshape minds (in pretending that we are working with a real brain without oversight), we must position ourselves as either heroes or villains. But how to choose? On the one hand, we can seem delusional or overly confident in our technical capacities, or simply unaware of what another person will need or find useful. We might initiate a change that fundamentally reshapes a life and not for the better. On the other hand, we have an opportunity to add to another person's experiences and participate in a much larger neuroscientific initiative: to map a universalizable brain holding so damn delicately neuronal connections propping up a "normal" life. We simply do not know what our probing, our insertion, our control will do in the end. We are forced to move forward. We answer the call of the surgeon. We must occupy that confounding space in *A-Me* even while the artwork refuses an easy answer to the dilemmas that it poses.

We can now wonder if Spinoza was way off-track when he said, "when a man is prey to his emotions, he is not his own master, but he's at the mercy of fortune."[33] Yet without the emotions—care, fear, anger, compassion—we could not consider what we do so thoroughly nor could we appropriately regard our actions in a project like *A-Me*. Perhaps Spinoza redeems himself in stating, "Desire is the actual essence of man."[34] No, on second thought, he merely splits the human in half—essence on the left, and logical "master" on the right. The damage of this kind of reason–emotion dualism is quite evident in *A-Me*. So we prefer, instead, the full effect of emotion on structures of thought. The embrace of the irrational desire and the passionate outburst spurs us to invent new thoughts more "rational" than that which flow from any disembodied head, which is, lest we forget, an image of violence.

Scan 6: the brain is magic

The brain organ as the contemporary arbitrator of the human (exposing our universal impulses, drives, limits, and capacities) becomes so very strange when feeling the weight of the thing in your hands. It is quite small. And the exterior of this wrinkled, grey organ is ugly, really. The brain's outer aesthetics do little to convince us that it is the Main Thing responsible for our vibrant experience. This is it? But when illuminated with magnetic resonance, colored in bright red and green, made to sparkle from the inside out like electric diamonds, the seduction of the organ returns. The magic is back. The idea of there being One definable able to explain us, able to rule us all, feels possible again.[35]

But what kind of magic is the brain? Traditional magic might be defined as a set of practices "by which one seeks to engage directly with external agents (including human, animal, and supernatural agents)."[36] The typical, and skeptical, view on magic proceeds as follows: gullible people displace their anxieties onto those with greater confidence, i.e. magicians, who may be living under self-delusion about the scope of their influence on the environment; magicians purport to illuminate hidden connections and thus persuade. The sharing and confirmation repeat, proving a trick to be real, perpetuating specific beliefs, which congeal over time into communal rituals wherein many accept that external agents are, in fact, altered by the actions of a magician. In this schema, the community eventually "demands conformity" to increase a felt spirit of power.[37] Magic at this stage invests in the psyche. It is a Becoming. The magician psychologically unifies the community and attunes to them, serving their needs.

Vegas "magic" makes for a good contrast. In Vegas, the skeptical mind withdraws from the scene for the purposes of the entertainment. The paranormal of David Copperfield, for example, explodes in light and smoke, coming alive for us within the suspension of disbelief. Copperfield visually stuns, excites awe, but yields no practical result for the audience, outside of a good night out to impress a date. The world seems altered, but only for a moment; the magic is never more than a trick. The performance remains communal, but the manipulations are technical, not to be too personal, too injurious, or too serious. The trick does not last nor permeate the psyche nor crawl into the core of what holds a community together. Compared to religious and cultic magicians, Copperfield is an abomination, out for prestige, not a dedicated servant resolving community problems through practiced exploration of Other Worlds. The Vegas magician is a performance artist subject to no accusation.

Here we can recall two defining features of magic: the exposure of the hidden and the expectation of control, of replication. The latter proves distinct from other religious acts; to replicate the act makes the imagined other-worldliness appear suddenly real, but the constant manipulability of it also has risks and affordances. The risk is that the audience sees the world as too easy, which would expose the 1-2-3 as a trick; the affordance, however, is that people might expect that they can control most things, at least to some extent, and those with special skills or powers can especially. As Bailey notes:

> the magician (like the scientist) assumed that certain actions properly performed would always produce identical results. Religion, on the other hand, emphasized the "propitiation and conciliation" of higher powers (See: Frazer, 1927, p. 48).[38] Priests might hope for certain results, and beseech the gods for them, but they could not expect their prayers always to be answered.[39]

Magicians stand confidently before "faithful practitioners and the broad understanding of the universe that they promoted."[40] In this way, the magic precedes the magic show. It lives inside of collective imaginaries from the human "expectation of control, of replication."[41] If the brain is magic, then it is magical out from this same formulation. Our images and expectations of the brain are its own ventriloquisms. If materiality ever surprises the neuroscientist who pursues exacting replication, then we might call the finding a fluke or a lucky accident, but we will never call it magic—because it was unintended, outside of the scientist's powers of creation.

Following this line a little further, neuroscientists (and neuro-artists) might also be said to practice a communal form of magic in setting out to expose "the private, secret, mysterious, and above all prohibited."[42] Probing inner thoughts and dreams, how they are made and why, is a communal exorcism, which works amid a demand for conformity—inside and outside the lab, in designations of normality and abnormality, in procedure and measurement. Neuroscience participates in a sort of magic making, at least in the conceptual sense outlined by anthropologists. In so doing, neuroscientists have a choice. They can either escape accusation by imposing rules against critical questioning and treat interpretation of data as access to higher (expertise) powers. Or they can operate within a zone of danger where the magic is not behind-the-scenes and where the community has input into the structured material arrangements that might one day demand conformity to the results that will affect them. The question for us, today, is whether we will clap and cheer for the show, be mesmerized and swept off our feet, or whether we will demand to see behind the curtain? What kind of attunement to the community will we require of this "magic"?

What I have proposed above is that neuroscience participates in an ancient social function also common to magic, and when taken seriously by the community and geared to solve problems, then it is not always out for its own gain. We can accept this characterization if also recognizing that science makes a concerted effort, in being science, to distinguish itself from traditional magic. In fact, to escape the marvels and lore of old magic and not have any reliance on "advantageous moments," scientists became such because they tested specific things and events as natural phenomena, participating in the development of the Enlightenment, contributing to the formation of core scientific values, such replication.[43] But that strong distinction—between a magic and a science—which needs to be made is precisely what makes the comparison and any underlying commonalities worth teasing out.

To say that neuroscience works for people in the same way that magic once did is to ask if neuroscience can, first, be careful enough with its own hype and detailed enough with its own interpretations to stand in a clear, strong contrast to the false representations of magicians who fool gullible suckers with concocted potions. Second, it is to ask if neuroscience can develop in respectful communion with the deep emotional investment that people have traditionally placed in local witch-doctors who live out, indeed perform with their bodies, a total investment in the Other in order to help them get through life. Can neuroscience take shape through intimacy with the particular social, psychological, and bodily sicknesses of the community? Can the magic inside the neurosciences be its communal spirit even when the magic of our own brains is the expectation of control and the recurring delusion that, we ourselves, can control more and more Things and should discover higher powers through the use of special tools?

To say that neuroscience is "magic" is to recognize its ability to excite and to create new realities for people in need, but more so, it is to offer a challenge—to see if neuroscience can attend to the close spaces of our bodies and communities. Likewise, to say that the brain is "magic" is to challenge neuroscience's audiences: to see if we can reverse the trick—to hold the expectation of non-repeatability and an unabashed lack of control over all the many variables and hidden folds of the universe.

Sculpting a brain: *Resonance Punctuated*

When thinking back to a historical materialist tradition trying to understand the formation of subjectivities through analyzing the conditions of prisons, workhouses, and factories,[44] the entire question of whether brains can connect to sensual experience and be creatively investigated through attention to environments points to one answer: "yes, of course." There is no hard and fast division of external material realities from internal emotional ones, which are also, lest we forget, unavoidably material. The sniff of a tire shop jogs a memory. The feel of a warm teacup relaxes us. The old wooden spoon given to us by our great grandmother is the image that we conjure when thinking of home. Sensuality is of the senses. If the mind is inextricably tied to the brain, then the brain is composed of the external world.

Laura Jacobson's sculptures remind us of this—the deep absorption of materiality in daily life. In *Resonance Punctuated*, Jacobson sculpts a series of brains from clay, cobbling together the regions and sub-segments by pressing in steel bits, wood casings, auto-parts, and computer circuit boards. Working from the visual designs evident in real fMRI brain scans, she highlights how everyday industrial settings alter neurobiology. Her sculptures open pondering about how people are shaped psychologically by things like tools and chemicals, little machinic routines. Accordingly, in Jacobson's art, how brain structures develop or what they might mean is not a domain exclusive to biology and not something solely fit for laboratory exploration; semiotic analysis need not take a back seat to neurons nor must social encounters and their emotional acupunctures be made irrelevant to the brain's "wired" connections. Such divides are erased in these tactile conglomerations.

For Jacobson, touching machines is touching the brain. The hammer and dust of the coal mine are part of our subjectivities. Yet, there is no shying away from the visual tension across the soft plasticity of some of the materials Jacobson incorporates, such the clay that she squishes into and around the hard rigidity of

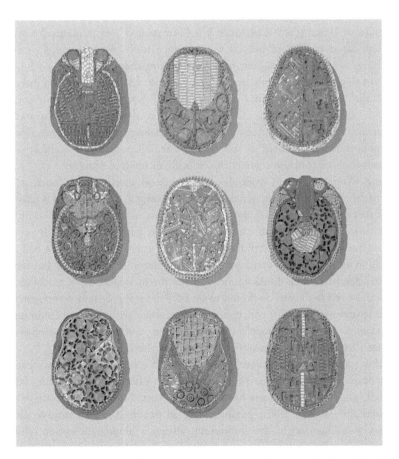

FIGURE 7 *Resonance Punctuated 2017*, ceramic mounted on wood 44 x 36 x 1.5in, Laura Jacobson, by permission.

other materials, like wire mesh. Made in this way, the brains, as artworks, imply that we are plastic to some extent, at some base level, but ordered with unavoidable determination by unsympathetic motors. It is difficult to view these sculptures and not walk away believing that we are, on some level, stuck; on another level, we see that we are deeply changed by what we do and the kinds of encounters that we have—the greasy, scrabbly work forced upon us by structural dependencies—which are pressed into us like clay.

Perhaps to strengthen the deterministic, Jacobson fires the clay. The brains are solidified into a ceramic piece. The representation is hard. The gesture can be metaphorical: brains, over time, turn solid like pots, ready to use, shaped to a purpose, but easy to break in half.

Or maybe Jacobson implies that brains, once captured by the scanner, can be organized on a shelf like pots. We can be arranged from normal to abnormal in a nice row. That is, if neuroscience is a kind of scientific hardening—making a picture of our mentalizing structure and, by correlation, of who we are and what we can do—then *Resonance Punctuated* declares that the external forces vibrating our Being (the "Resonance") is "Punctuated" (or punctured) not only by tough labor conditions but by neuroscience itself. And maybe she's on to something: we can "harden" our minds with neuroscience.

As sociologist Joseph Dumit argues, many people adopt neuroscience categories to describe and understand the Self.[45] Jacobson's sculptures remind us of this. However, there is another way to see the function of neuroscience: like molding clay, neuroscience takes the shape of a bowl, meaning that neuroscience feeds us; it is comfort food. We hold its outputs, and the shape provides comfort within the production of concrete representations with an obvious purpose.

This is where Jacobson's work also critiques neuroscience just as much as celebrates it. In fact, looking at her sculptures, we might be compelled to say that neuroscience shows us nothing about our brains that we do not already know from folk knowledge and common experience. Grandpa who worked for fifty years as a farmer thinks in terms of fertilizers, tractors, and seasons; he rises from bed automatically with the sun and cares not much for politics. Grandma who has for fifty years managed an office thinks about life in terms of ledgers, copy machines, and invoices; she sets her alarm clock, just to be safe, and is highly skilled at negotiation. The structures of mind have conceptual resources built with the lived environment. Neuroscience will not do much more than indicate, sometimes more precisely and sometimes less, where those areas of the brain are located.

Looking at *Resonance Punctuated* we can also be more positive about neuroscience's potentials. We can see the extent of our plasticity, how much clay we have left, and where the gears have started to rust. Neuroscience bears witness to materiality—it is a realist pursuit, after all—even if there's always another manifestation lingering in the scaffolding, always a hidden dimension. But the sculptures do seem to make us wonder if those steel bits can be yanked out of the fired pot and if more clay can be added. Can we change, or does a lifetime in the factory prove intractable? Maybe neuroscientists can be our mechanics?

Laid out as a collection, the brains look strikingly similar in their tonality and construction but quite different in the configuration of the details. Some are packed with metallic circuit boards, others lined with wood and tiny gears. From top to bottom, the artist foregrounds how we are shaped by what we do, implying an interior–exterior nature of brains. Perhaps we are individualized, little bit by little bit. That is to argue that the dominance of industrial and technological gear across all of the brain sculptures foregrounds the external influences on how we think. Whatever plasticity was once possible finds definition in our exterior activities; we remember our reliance on motorization and automation and computation and then, broadly, we see our technological co-evolution.

In some ways, the artwork is a worry about technology. Even if the clay highlights a connection to earth, the mucky stuff is hidden and trapped in-between grinding metal. The Biblical story of God breathing life into the dust of the ground to create humanity looks suffocated by an obsession with technological progress.[46] The tight circumference of Jacobson's ceramic skull is overloaded. The pursuit of god-like capacities through technological extensions transform the mud of brains into machines. *Resonance Punctuated* visualizes an old Frankenstein story. The original clay of the brain cannot at this juncture, we imagine looking at these works, escape its ruddy techno-configurations. The brain is too jammed up and scarred with human cyber trash. We are machinic now.

But reading the sculptures either as the image of the concretizing influences of neuroscience or as a picture of a human unable to be un-technologized is a narrow view. The neuroscience scans used to guide the making of these sculptures do not need to reference the specific neuroscience experiment; the brain pictured here might be broadened out to refer conceptually to a brain made over extremely long periods of time out in the world. If the sculpture is explicitly referential and, in that sense, a realist depiction, then the object referenced is not merely a scan. It is a whole scope of environmental influences and evolutionary adaptations. The brain takes shape from an evolutionary history much older than any neuroscience image. The brain is only as automated as automation, as regulated as regulation, as hierarchical as the oldest hierarchical structures. The impulses of humanity, its time-tested associations and social configurations, its sensations and journeys, its food sources and animal allegiances, likely make a much bigger difference than the contemporary life of the video gamer or the discourses of digital dominance or the neurograndiosity now pervading pop culture magazines. Who is to blame, then, for a brain that thinks like a machine?

The brain image that we see is not only referencing a particular brain scan, an exterior world of working environments, or an ecological deep history, but it is also recalling a history of art concerned with the corruptive and persuasive influence of mass-produced materials. Jacobson's work, perhaps necessarily, harkens back to 1960s Nouveau Réalisme. A troupe of French artists, including Arman and Jean Tinguely, "set out to prove the death of art's preciousness by considering reality their primary medium" and combining found and discarded objects.[47] Trash works satirized the clean, staid, careful comportments of high art just as much as consumerism and overproduction.[48] The attempt to bring back down to earth the elevated and idealized of the neurosciences seems to recuperate the strategy.

Jacobson's work offers something like a secret history of a brain without looking directly at brains themselves, i.e. without being obsessed with flesh and neurons as such. The brain is, instead, what it emerges out of, and in this way, the artworks fit the emphasis on the *everyday real* defining the Nouveau Réalisme movement. The first manifesto of the Nouveau Réalisme movement states, "If one succeeds at reintegrating oneself with the real, one achieves transcendence, which is emotion, sentiment, and finally poetry."[49] In contrast to a neuroscientific transcendence pursuing a smarter and more agile brain through neuronal reorganizations and neuro-pharmaceutical adaptations, Nouveau Réalisme's engagement with "real" materiality works not on the level of intellectual heights but of emotional depths— artists like Arman seek an emotional transcendence, i.e. making use of old things to separate from the needs and pulls of consumer delights. So there is an inherent tension between the body's feelings about materials and the capacities of the material itself, how Things tug at bodies to live and act differently.

Believing that creative display of an industrial "real" can jar the interior seems a fitting way to understand Jacobson's sculptures, precisely because she crafts an image of the neurobiological "real" with objects normally thought to be exterior to it. And the effect is similar: we gain some emotional distance from the neuroscience lab; we see in her sculpture a kind of transcendence about how brains function (or take shape and shape us). And we can then see life as poetry, or maybe as a terribly composed verse, bellowing out a tune all about, no, *composed of* America's industrial wasteland.

But Jacobson's brain sculptures ask another question of us, one about our reactions: is the brain too concretized to be altered by witty art interrogations? This is a recursive, self-implicating question. If drawing from a history of Nouveau Réalisme, then Jacobson likely doubts the optimism of an art enterprise being a means to freedom. We see this when we she makes her sculptures so hard, chunky, and tooled. Perhaps, she turns Nouveau Réalisme on its head by offering us a sardonic critique of its fanciful aim to set people free from mass produced consumer culture.

Indeed, staring at *Resonance Punctuated*, we are not confident that we can change. The artist's attention to environmental co-being pushes an index finger into our own clay—to see how flexible we might still be. Like many of the best sculptures, the shapes suggest something that we already *feel* deep down inside: a world pressing in on us. Maybe we are reminded of our mobile phones. A need for increasing alterations or the need to keep up with social media. More speed. More efficiency. We drown in fears that we are all swimming in "The Shallows," to recall Nicholas Carr's extraordinary warning that digital media is changing our brains and not for the better.[50] Jacobson's sculptures can be read in this way, but the process of change is really quite old, as old as the first day on the job—no much, much older, as old as the thumb on the chimpanzee. Yet we somehow do not feel comforted by the fact that we have a thumb when we see brains like *Resonance Punctuated*.

Look at the yellow line bisecting one brain, then the blue wave across another, then the burnt red char, then the linear composition, then the wheely twirl of the sprocket. *Resonance Punctuated* suggests that we are a mess. But a beautiful one, kind of. We are a compacted garbage container of environmental techno-trash whose skull can barely hold it all together. But the life that shapes it all eventually congeals into structures and sub-structures that give us some pleasure, even if the material agencies there take us for a ride sometimes. If *Resonance Punctuated* "works," then the response of the viewer is probably "yeah, I *cannot resist* my screen for like more than two seconds!" Or, "I can't help but see *everything* as a movie." Or, "that image would sure look better with my new saturation filter."

The impact of Jacobson's material forms—amid our knowledge of what copper and steel edges feel like—connect the neurosciences to our everyday experience. And they sing a melancholy tune alongside a rising chorus about how we are all changing, and much too fast. *Resonance Punctuated* in this respect provides a necessary biopsy, an open-heart brain surgery to diagnose the clogs in our thinking about what we can do if we feel a need to change our brains. But what Jacobson finds when she cracks open brains are local environments—and there's nothing really very scary in there at all, just the old machines that we know and love. Thus, if we hope to change, then we must clean things up a bit; we must talk to the bosses who decide what our everyday involves, or we must relocate to new landscapes. But whatever we do, Jacobson's work is certainly something that we feel—an embodiment negotiated day by day, a composition of material structures where brains become car lots, warehouses, and computer stores.

Scan 7: the brain is a way forward

The social and philosophical history of liberal humanism—a staging of rationality as the defining factor for human Being—scaffolds the persuasive capacities of the neurosciences. Logically outlining how humans think by mapping the brain networks of every conceivable experience and calculating interconnections will not dampen much the notion that rationality defines us. Indeed, science's approach to an organ, pre-understood as operating from "systems" and "computations" in a way that will ultimately "makes sense," automatically embraces experimental rationalizations. Adopting the regimented disciplinary discourse might presume neuroscience is inseparable from liberal humanism's core tenets (the superiority and independence of the human in addition to the production of objective and unassailable truths). However, there is a surprise in the making. The rejection of liberal humanist ideals that does not allow rationality to be set free from emotive life nor logic to be independent from social structures can also, in an unexpected kind of reversal, strengthen the epistemological position of the neurosciences and enhance the need for it.

A renewed investment in emergence and dynamism striving for an interdisciplinary reunion of mind and matter in academic scholarship (which we can call "New Materialism")[51] allows brain images to be understood as cartographies of the unconscious and as charts of developmental emergences. That is to say, brain images provide access to the tiniest particularities of human Becoming. The neurosciences not only outline how we think in terms of making judgements but also how bodies transform with the suasions of emotive life, as appeals to emotional circuitry and to complex "processing" are folded into the New Materialist picture. The same can be said for sense perception and unconscious awareness of environmental cues. So the neurosciences exercise, and thus retain, high epistemological status both when liberal humanism reigns supreme as well as when bodies and environments are staged to take back some agency. Whether the brain is understood as a solitary, rational general dictating human existence or, conversely, as integrated set of co-forming networks performing human experience in real time, we can still approach basic questions of human life through the sciences of the brain.

Regardless of whether we privilege the interior or the exterior, the rational or the emotive, or some blend, in thinking about the human, the brain stakes out an unavoidable claim to the explanatory means. Recent advances only add to the persuasive case. As LaBar notes, "Emerging technologies, such as optogenetics, permit an inquiry into brain function at unprecedented levels of detail and sophistication."[52] The brain has never been seen in more detail than today, and technological capabilities only improve. "From epigenetics to social neuroscience, scientific endeavors are leading the way to a new, integrative understanding of the neural bases of complex disorders and mental states. These advances in neuroscience herald a promising era for translating basic knowledge into practical tools to improve the human condition."[53] Who could argue with the pragmatic impulse to "translate" neuroscience into "tools" for living?

The utility of neuroscience has not been lost on non-expert audiences. Rarely do I go to a dinner party and explain that I study the social effects of neuroscience without hearing someone say something along the lines of: "Oh, I hear that superfoods are good for the brain!" or "Check out this brain game on my phone! I'm using it to improve my memory." As Thornton notes with some humor and some disdain, neuroscience organizes many discourses of the body and promises radical new self-understanding. Neuroscience is popularly positioned as "an accessible body of knowledge" hyped in self-help books and applied to nearly every aspect of human life.[54] Findings then become uncritically adopted into personal vocabularies. Articles such as how to know whether or not you have a "normal" sex drive,[55] or "How your brain is affected if you skip the gym"[56] help resolve everyday problems. The language of neuroscience settles common concerns.

Extension of scientific findings into lifestyles product marketing may well be the norm in a Capitalist society, but brain science, perhaps more than any other science, gets geared up to solve a multitude of problems. And the ridiculousness of many claims demonstrates just how useful it is presumed to be. This popularity, as Thornton explains, exists because the brain is understood as "the source of literally all human thought, emotion, and behavior" and as a result, marketers suggest that "willful efforts to improve the brain will naturally lead to superior intelligence, greater emotional stability, and improved performance in the home, in the gym, and at the workplace."[57] A similar observation was made by Nikolas Rose. He argues that psycho-pharmaceutical interventions are now privileged in the West over folk remedies, lifestyle changes, or environmental adjustments, and many are prone to think of psychological conditions first as chemical imbalances.[58] Now more than ever, beliefs and feelings—the love of our partners or of chocolate ice-cream or of chess or of Facebook—are set in terms provided by neuroscience and biomedical researchers. It is fairly common to hear people talking about their brain and efforts to "rewire" themselves, deciphering which neurons are "plastic" and which are static. Living better and longer is a matter of knowing more about the brain.

The brain in neuroscience headlines leads us back to the heart: we adopt the neuro-craze out of hope for a positive way forward through the brain. We want self-understanding. And we also want to live with ease. We do not want to go haphazardly along. We need to monitor how we are changing, and we require at least some agreement around basic forms of knowledge to do so. What can we do, then, but attend to the electrical and chemical life of the brain? It makes no sense to ignore it.

The brain offers a way forward in an age of uncertainty. Amid a messy future where we must breathe in over-crowded urban environments, must grow food amongst soil toxins, must confront the consequences of trash-laden oceans, must negotiate a terrible lack of energy resources, must swallow bloated arguments from politicians, and must constantly decry "fake news" streaming through our mobile phones, it sure would be nice to have some means to care for strange psychologies, to settle emotional turmoil, to resolve addictions, to monitor ugly tendencies, and to improve political decision-making. We need a solution right close to us, some place to find answers, some means to do something for ourselves. We need a brain. And at the present moment, we need a neuroscientific one at that. But this should not keep us from questioning liberal humanist ideals nor from noting how history lingers in our practices and how we might improve upon our approaches.

Thinking neurosensuality's New Materialism

Taylor, Puig and colleagues, as well as Jacobson present diverse artworks that, in one way or another, investigate the overlap between neuroscience, technicity, and sensuality. Their cross-disciplinary engagements seek closer attention to inner thoughts and feelings while the materiality of each work turns the audience outward to the earthly forces of human development. Collectively, these artists expose the palpable tension between the agency of things and the constructed Self, and they do so at a time when both affect and materiality arise together to amend the semiotic obsessions of the last few decades. What, then, of neurosensuality's tie to New Materialism's goal of compensating for the limitations of post-structuralism?

Thinking about New Materialism as a broad-based effort to dismantle old disciplinary boundaries and an attempt to unify biology and sociality, neuro-sensual brain art—as a ground-up, heart-felt attempt to enact tenets of New Materialism—discloses the immensity of the challenge. Put simply, there are at least three difficulties facing New Materialism's dreams that can be also described as three risks evident in neuro-sensual brain art.

The first is pursuing scientific materiality by moving outside of the science's own circumscribed bounds of disciplinary legitimation. Neuro-sensual brain art does this most obviously through the sentimentalization of scientific practices. Sentimentalization puts the objectivity of science under the microscope. Yet, performing scientific processes and bodily engagements in spaces familiar to humanities inquiry—like art galleries or theaters—presents an image of separate domains. The sciences may appear unable to comment much on the human processes of their own laboratories and so are left to do normal business, unbothered. Likewise, perhaps, the humanities risks looking like it does not fit in nor into the laboratory. When the artist leaves out, or sets aside, the specific methods and locales of the neurosciences, as in several cases examined here, the creative representation risks appearing exterior to disciplinary boundaries and can more easily be delegitimized. The artwork may also seem added-on after the fact, not ever intended as part and parcel of the making of brain science. And art—why the hell not?—can be right in on the act.

Neuro-sensual brain art may also risk implying, at least at moments, that scientific activity lacks the emotional sensitivity or the ethical relations to fully consider itself. Neurosensuality seems to gain most of its exigence from bringing a more "human" lens to the brain sciences. The artist can be too easily positioned as the humanist hero there to save the scientist from gross disconnection and misappropriation. Yet, New Materialism's assertion that the humanities must adopt analyses inclusive of nonhuman materiality seems to suggest that no individual's perspective fully captures material events; so New Materialism's turn back to bodies, affects, and feelings, generally, must be attenuated with a view that ethical concerns are also enwrapped in large networks and constrained by institutional regimes including charts, optics, and the magnetic properties of scanners. In brief, there is a tension between what is suggested when artists are said to be needed in neuroscience labs and what is suggested when New Materialism foregrounds support structures. New Materialism aims to complicate any humanistic dominion over what humans (might, do, or could) see and feel; yet when it comes to making sensual, biologically informed art, the focus on lived histories and phenomeno-logical reflection risks overwhelming the call to be a more-than-human network.

But there's a third downside as well, which is waiting to pop up like a troll crouched on the opposite approach. Neurosensuality risks subjecting emotional response to the strict domain of data and measurement, introducing stultifying order, normality, and hierarchy quite unusual when a taste of the strange, of generativity, and of multiplicity are preferred for humanistic and artistic inquiry. The dimensions of personal experience may be squashed down by neuroscience discourses and neuronal aesthetics. This point is more apropos, perhaps, for the next chapter; but in any event, the artist adopts the neuroscientific lens to get closer to the real and to touch what is experienced—but the result may, at times, at least, be flat obedience to data over and above ironical criticism. How-ever, neither is entirely desirable on its own.

Making art with neuroscience is a tricky affair. Over-sensualizing the scientific, as side-effect of interdisciplinary exchange, risks sappy art and tenuous science; under-sensualization risks downplaying the power of art to affect us and overplay-ing the power of science to define us. In this way, innovative brain art projects, like many interdisciplinary hybrid New Materialist projects, confront a tough impasse. The barrier is a historical set of circumstances governing disciplinary action and expectation. The politics of authorization and legitimization are assert-ive on both sides.

As a result, science's pragmatic dedication to delineation and definition can go completely unaltered by art's opposite inclination to open-up everything for fur-ther exploration and emotional evaluation. Reading the brain art of this chapter, indeed, foregrounds how the works operate outside of the appropriated domain, even if they do push some buttons. It remains unclear if artworks that interrogate methods or intrude on researchers' respective theoretical commitments, as Puig and colleagues' *A-Me* seems to want to do, will have any long-term effect.

As of now, the inclination to restage neuroscience as meditative practice or to use it for confessional purposes balances uncertainly between reifying and reforming neuroscience. On the one hand, neurosensuality creates an emotional stage where doctors and researchers discover new connections between neurology and their patients' everyday lives. Thus, this kind of brain art helps to overcome any ten-dency to view the brain as sheer data open to manipulation and chemical alteration absent consequences that resonate across many bodies. On the other hand, neuro-sensuality in brain art can leave unaddressed what neuroscience does on a broader, longer scale to revise its interventions in lieu of patient experience. Heightened individuality and emotionality in the works tend to prioritize the here and now, and they also tend to centralize specific people's identity formation. In brief, con-tributing to emotional awareness while also addressing structural conditions without coming off as a humanist savior or a critical jackass is a difficult charge.

Few works will be able to rouse lasting impressions while simultaneously having the right timing and the means to thoughtfully contribute to the neuro-sciences. The trick will not be as simple as putting the neuroscientist into the artist's shoes or vice versa. If neuroscience can absorb or ignore an artwork, take its lessons to heart or laugh at its interventions, then artists probably cannot do the same in equal measure. Artists cannot expect to bring aboard the most serious of the disciplinary regimentalists without some expression of respect for the practice and the real of scientific materiality being inscribed. Then again, being loved and taken seriously while being constructively lampooned is a classic rhetorical strategy for family members to express a desire for behavioral change. But artists and scientists have to feel like family first.

Notes and references

1 Eric Racine, Ofek Bar-Ilad, and Judy Illes, "fMRI in the public eye," *Nature Reviews Neuroscience* 6, no. 2 (2006): 159–164.
2 Walter T. Stace, *Mysticism and Philosophy* (London: Macmillan, 1960), 85.
3 Ibid.
4 The notion of "extended mind" comes from Andy Clark. See Andy Clark, *Natural Born Cyborgs: Minds, Technologies, and the Future of Human Intelligence* (Oxford: Oxford University Press, 2003).
5 John Paul Sartre, *Existentialism is a Humanism*, translated by Carol Macomber (New Haven: Yale University Press, 2007), 17.
6 Ibid.
7 Daniel C. Howe and John Cayley, "Reading, writing, resisting: literary appropriation in The Reader's Project," in *Proceedings of the 19th International Symposium of Electronic Art, ISEA2013*. University of Sydney, Australia, 2013. Available at: http://thereadersproject. org; Daniel Howe and Braxton A. Soderman, "Generative art and the aesthetics of surprise," in *Leonardo Electronic Almanac 24.1* (Cambridge, MA: MIT Press, 2017); Jhave David Johnson, "Rerites (poems written by neural nets)," *Glia.ca*. Available at: http://glia. ca/2017/rerites/.
8 David Wills says that Nietzsche wanted the reader to understand the message "God is dead" as "old news" just as Zarathustra, the lead character in Nietzsche's book *Thus Spoke Zarathustra*, seemed to have trouble (like Nietzsche) accepting that anyone really believed that it was true and, thus, really old news. See David Wills, *Dorsality: Thinking back through Technology and Politics* (Minneapolis: University of Minnesota Press, 2008), 210–212. Here, accordingly, I use the phrase "old news" to signal the more pressing news that God lives as in another form, most pertinent here for the discussion in this book in the form of a brain, as The Brain. The search for comfort is directed there. Thus, the "new news" is the news that humans are never able to abandon God, in one form or another. But once situated today, the new news is how much a stressful new media environment makes this "old news" no longer useful news at all.
9 Sartre, *Existentialism*, 22.
10 Alex Williams, "How anxiety became society's prevailing condition," *Independent*, June 17, 2017.
11 "Vasily Kandinski (1866–1944)," *Guggenheim*. Available at: www.guggenheim.org/ arts-curriculum/topic/vasily-kandinsky-composition-8.
12 Daniel Robbins, "Vasily Kandinski: abstraction and image," *Art Journal* 22, no. 3 (1963): 146.
13 Marjorie Taylor, "Warm Glow, or Fabric MRI: Bill's Brain," *Museum of Scientifically Accurate Fabric Brain Art* (2009). Available at: https://harbaugh.uoregon.edu/Brain/.
14 Deena Skolnick Wesiberg, Frank C. Keil, Joshua Goodstein, Elizabeth Rawson, and Jeremy R. Gray, "The seductive allure of neuroscience explanations," *Journal of Cognitive Neuroscience* 20, no. 3 (2008): 470–477.

15 Beth Ann Pentney, "Feminism, activism, and knitting: are the fibre arts a viable mode for feminist political action?" *ThirdSpace, a Journal of Feminist Theory and Culture* 8, no. 1 (2008): para. 1.

16 Ibid., paras 2–4.

17 Beryl Tsang, "Titbits," *Knitty* 13, Fall issue, September 12, 2005. Available at: http://knitty.com/ISSUEfall05/PATTbits.html.

18 Pentney, "Feminism," para. 20.

19 See Leah Rumack, "Cozy cleavage," *The Globe and Mail*, October 15, 2005. Available at: www.theglobeandmail.com/life/cozy-cleavage/article18250809/.

20 Pentney, "Feminism," para. 1.

21 For discussion of fabric art's sometimes odd and lowly status compared to other arts, see Janneken Smucker, "Paradoxical objects: quilts in American culture," *Perspective* 2, online (2015): paras 9 and 25. Available at: https://journals.openedition.org/perspective/6076.

22 See Gordon D. Kaufman, "Theology, the arts, and theological education," *Theological Education* 31, no. 1 (1994): 16–18.

23 Michael Brookes, "How to knit a brain," *New Scientist*, December 17, 2008. Available at: www.newscientist.com/article/mg20026873-200-how-to-knit-a-brain/.

24 In this passage, I am inherently recalling a theory of Multiple Ontologies as detailed in Science and Technology Studies. See Annemarie Mol, *The Body Multiple: Ontology in Medical Practice* (Durham, NC: Duke University Press, 2002).

25 Jordi Puig, Andrew Perkis, Aud Sissel Hoel, and Alvaro Cassinelli, "A-me: augmented memories," *SIGGRAPH Asia 2013, Hong Kong*, November 19–22, 2013.

26 Jordi Puig, Anamaria Carusi, Alvaro Cassinelli, Andrew Perkis, and Aud Sissel Hoel, "A-me and BrainCloud: art-science interrogations of localization in neuroscience," *Leonardo* (2016): 6–9. Available at: http://eprints.whiterose.ac.uk/110174/1/Puig_02.pdf.

27 Puig et al., "A-me: augmented," 1.

28 Ibid., 2.

29 See Chapter 1 of this book.

30 R. Lanier Anderson, "Friedrich Nietzsche," *Stanford Encyclopedia of Philosophy* (March 17, 2017): 6.1. Available at: https://plato.stanford.edu/entries/nietzsche/.

31 Michel Foucault, *History of Sexuality*, Vol. I (New York: Vintage Books, 1990), 140.

32 Katia Genel, "The question of biopower: Foucault and Agamben," *Rethinking Marxism* 18, no. 1 (2006): 44.

33 Benedictus de Spinoza, *The Ethics* (The Project Gutenberg Ebook, 2007), 89.

34 Ibid., 88.

35 *The Lord of the Rings* reference is intended. See J. R. R. Tolkien, *The Lord of the Rings*, 50th anniversary edition (New York: HarperCollins, 2004).

36 Gregory J. Wrightman, *The Origins of Religion in the Paleolithic* (Boulder: Rowman & Littlefield, 2014), 121.

37 Ibid.

38 James George Frazer, *The Golden Bough* (New York: Macmillan, 1927), 48–51.

39 Michael D. Bailey, "The meanings of magic," *Magic, Ritual and Witchcraft* 1, no. 1 (2006): 3.

40 Ibid.

41 Ibid.

42 Ibid.

43 Steven P. Marrone, *A History of Science, Magic and Belief: From Medieval to Early Modern Europe* (London: Macmillan International Higher Education, 2014), 93.

44 See James A. Aune, "An historical materialist theory of rhetoric," *American Communication Journal* 6, no. 4 (2003): 1–17. Available at: http://ac-journal.org/journal/vol6/iss4/iss4/mcmcgee/aune.pdf.

45 See Joseph Dumit, *Picturing Personhood* (Princeton: Princeton University Press, 2004), 1–6.

46 See Genesis 2:7, *The Berean Study Bible*. Available at: https://biblehub.com/genesis/2-7.htm.

47 "Nouveau Réalisme Nouveau Réalisme," *The Art Story*, para. 1. Available at: www.theartstory.org/movement-nouveau-realisme.htm.

48 See "Documents of Nouveau Réalisme collection at the Menil Collection," *Niki de Saint Phalle*, March 20, 2010. Available at: http://nikidesaintphalle.org/documents-nouveau-realist-performance-menil-collection/.

49 Ibid., para 2.

50 Nicholas Carr, *The Shallows* (New York: W. W. Norton, 2010).

51 See Diane Coole and Samantha Frost, *New Materialisms: Ontology, Agency and Politics* (Durham, NC: Duke University Press, 2010), 1–6.

52 Ken LeBar, "Advances in neuroscience," *Science Advances* 3, no. 11 (2017): eaar2953.

53 Ibid.

54 Davi Johnson Thornton, *Brain Culture: Neuroscience and the Popular Media* (New Brunswick: Rutgers University Press, 2011), 2.

55 Sophie Borland, "Is this why some women don't like to make love? Scientists discover those with low libidos behave differently," *Mail Online*, October 26, 2010.

56 Anna Lewis. "Here's what happens to your brain when you take a 10 day break from exercise," *Cosmopolitan*, September 1, 2016.

57 Johnson Thornton, *Brain*, 2.

58 Nikolas Rose, "Neurochemical selves," *Society* 41, no. 1 (2003): 46.

3
AFFECTIVE REALISM AND BRAIN ART

As with other art realisms, Affective Realism expresses a long-standing desire to touch "reality" in the singular, to achieve some clarity of the sensible, and to represent an experience with visual exactitude. Like realist predecessors, spotlighting the mundane, the overlooked, and the microscopic means that the hayfork or the glass bottle or the eyelash become symbols of an undeniable Truth. Accessing an ultimate "real" generates the wonderful compulsion to get so close to the shape and color and smell of the thing as to confuse the eye; the realist trick that tells the mind "this is a pipe" pushes right to the edge of what representation can do—fool the viewer into seeing what is not there and then return the question: is it a pipe? Is any Thing accounted for, really?

Metaphorically, Affective Realism is the over-the-top pop science headline, "This is Your Brain on God."[1] The headline is the garish highway sign on the way to Vegas, a giant flashing arrow, which calls the driver's attention to a cheap diner with slot machines and hookers near the restrooms. "Here lies the strangest and most intimate of your experiences" in bright red and blue. Come inside. Try the hash browns. They're only $1.99!

Affective Realism is a theme popping up within and unifying the brain artworks explored in this chapter. It can be described as an effort to discover a transparent and objective exposure of secret human truths about the emotional life, often focused on sudden feelings or affective experiences.[2] Works swimming in the crisp, refreshing colors of the Affective Real claim to outline—universalizable and total—our feelings on graph paper, and any aesthetic pleasure is automatic, built into the revelation. These artworks curl a finger toward themselves ("come over here and lick us up") like a Ralph Goings image of donut, offering a cold hard cortex with the promise of a secret inner pleasure state.[3]

Goings is a useful foil. His paintings of mundane objects from middle America help crystallize Affective Realism by setting it, appropriately if not conveniently, in relation to a cut-rate diner. The diner, Goings shows us, is a place for America to feed, a place where we can go to understand our strange little universe. We can dip our French fries into the ketchup and watch the brain glow with delight. But Goings was a contextualist. He was an interrogator of human experience through objects, and he revealed life only indirectly, working through the placement of the ketchup and the empty green booths. Nothing so secondary could serve as commentary on our existence in the Affective Realism of brain art. We do not discover the placement of glass beakers or rows of dark, empty fMRI machines. Affective Realism is much more about the exactitude of visualizing neurons as states of Being, of crafting visual performances of the most real *ever* seen. If Goings painted the donut, then Affective Realism would paint the salivating brain.

To play out the Goings comparison a bit further, we should note that Goings tries to disturb us. We get the opposite feeling most of the time from Affective Realism. Goings makes lovely photorealism because copying a photograph in his time "was a bad thing to do ... I was delighted to be doing something that was really upsetting people."[4] If there is anything upsetting about the material real in Affective Realism, then it is only sensed after the visit to the museum embeds itself under your nighttime blanket. There, once down on your pillow in the dark, you wonder how the artist could so nakedly put people's neurological disorders and dispositions up on display; you wonder why and then get a chill when you question whether that person—that brain—was really you. *Oh no, it couldn't be!*

Nothing immediately shocking slaps the viewer across the face when enjoying the bling of a delightfully colored brain scan. This kind of art is non-offensive. The exposure of underlying mechanisms is, indeed, a social duty, it can seem anyway; it is a good thing to do. Today's neuro-pop congregation want more, and they are giddy. The fright of the secret alien brain implant, the sick institutional domination of lobotomy, the inescapable Big Brother of mind reading, if anywhere, swims deep, deep down in a psychological well.

Affective Realism, as another art realism, pretends to create the perfect version of the other side of Chuck Close's *April*.[5] It is the raw interior of that perplexing face. In Close's image, the girl we take to be April is caught in heavy suspension. Like so many classical works of art, she flashes a look somewhere between defeat and desire. Or maybe, we think, her face venerates the artist. Or is it a veneration for us? She looks right at us. Are we the ones held in suspension? We dare not be tempted to simplify her extraordinary psychological portrait by saying, "Hey, light up April's brain and hang it next to her in the museum!" To go down this road would be to embrace the Affective Real, to try to make the "more real" of art the technical scientific explanation of our reactions to it. To see the emotionally jarring representation of the body as not "really real" is a tragedy that Affective Realism never notices in its narrowed gaze. It only sees The Brain.

Scan 8: the brain is a phoenix

The networked, neuronal brain rises like a phoenix amid the ashes of ignorance about economic and technological complexity. A tangle of socio-political unrest serves as the pitch-black background for the flaming beast to soar. The brain is of unmatched beauty in an age where questions about how bodies integrate environments and change over time are frustrated; these questions cannot be resolved by the super-massive ocean of information streaming in Wi-Fi signals all around our heads.

How unsettling it is to not find an answer on Google these days. As of yet, millions of online discussion forums and webpages have been unable to resolve basic questions of humanity. Perhaps we are too dumb to make the connections among pages ourselves. Perhaps history is too thick with competing ideas. The benefits of ecological perspectives are frozen in global-scale problems and superstructures too complex and dispersed to grasp.

Here, in a wash of the undecidable, we limit our scope and imagine ourselves as walking brains. In fact, if we do Google questions of our existence, then Google itself becomes the first problem that we must solve as brains. Said simply: we discover (on Google) that our brains are changing as a result of the immensity of the new media ocean, and we are already drowning. According to the narrative, our ideas, as well as our interactions, are only becoming "shallower" as a result of constant connectivity in an ever-expanding network.[6] The answer that new media provides to a condition where we feel the need to Google the answer is one about the trajectory of the brain tumbling into degradation, insensitivity, and aloneness.[7] The human-as-brain in a technological environment risks slipping deeper into disillusionment and ignorance about its conditions.

Yet, in another reversal, the brain emerges (from the ashes, a phoenix!) to solve the problem of living in a wild environment. Taking control of ourselves-as-brains, with and through our technologies, retools us for excellence. If viewed as a computational machine, then the brain—should we say "us"?—needs only to be reprogrammed. Right? We can keep our Google and Excel if we make ourselves into compatible programs.

To achieve the task, we will need to manipulate human neurobiology and gather as many empirical correlations as possible. We are well on the way. Put a person in an fMRI, and make her look at an image of a loved one. *Look at the neurons firing; they are bright red! These must be love neurons. Here is the region of the brain responsible for processing feelings of love.* By looking at these images, we purport to know what love is or, at least, most basically is. The ontology of the brain—its neuronal, synaptic processes—thus grounds an epistemology.

The persuasive effect can be strengthened significantly when the brain scan is mistaken as a body photograph. Roskies calls this "the naïve public consumption of brain images," i.e. understanding brain images in the same way as photographs used in historical investigations or in the courts.[8] Presuming mechanistic production and epistemic solidity in scientific imaging allows audiences to eliminate an underlying suspicion of ideological conjecture or politicization in the making of the image. The brain image can then be distanced from fears that its interpretation is rhetorical. But if the image of a brain is at all shaped by the negotiated parameters of the experiment, crafted by software packages, described through relationships to other studies, then it is rhetorical through and through. The footing is found in correspondences and similarities to past studies and related theories, in no way independent. Trust at that point, however, grows uneasy; the knowledge made and the meaning asserted is more easily opened to revision. Yet when the brain image is understood as an automatic production it stands as itself and becomes a more secure representation providing access to the "really real."

The more we turn to brain images to learn how to overcome new media's numbing dispersals or learn whether we really love Andrew or not, the more neuroscience intervenes at the very heart of philosophy: what is real, and how do we know?

The result is that philosophy, along with sociology, rhetoric, media studies, and cultural studies, cannot ignore neuroscience because of repeated claims about what is truly real, or what's more real. The popular neuroscience articles that regularly proclaim that "Baby brain really does exist,"[9] or "Too much TV really does rot your brain"[10] make the point. In these headlines, we find mechanisms to affirm what we already think or have long suspected. Folk knowledge finds concretization in the "hard data" of the scan. The discussion of "the real" is not ultimately decided by intuition or emotional awareness nor is it secured in community agreement or sociological observation—it is achieved by the protocols of neuroscientific validation. Yet any unassailable confirmation of "the real" is a deception but often only a self-deception, not a scientific one; science is, we should remember, a long, tiring process of revision and also a matter of probabilities.

The neurosciences have their own folk knowledges, logics, emotional drives, and communities to contend with. As a practice, neuroscience is not freely accessing an objective body; we should blame ourselves if we think so, because something within us—curiosity about the body or a desire for quick endorsement—wants the uncomplicated. The "neuro realism" embedded in popular news headlines that recycles generic social constructs ("Baby Brain" and "Bad Teenagers") is often missed, even if it obviously relies upon social presumptions. Consider those "Bad Teenagers" watching TV; the headline circulates the old belief that TV shows rot the brain.[11] But we often miss or ignore the cues that expose the social. It should be easier for us to see that no one in a study on the effect of TV shows, for example, is testing how TV viewing might advance cultural knowledge, build sociability skills, improve motivation, enhance verbal wittiness, or increase other mental capacities that we could consider positive. We have no one to blame but ourselves for uncritical neuro-fascination. But I want to argue that these cases may be mere side-effects of a broader impulse: the deep need for answers to complicated social questions, and more so, the hope of answering much bigger questions about life than those addressed in the studies. What shall we make of ourselves? How shall we then live?

We are reminded that neuroscience cannot be articulated without drawing on—intentionally or unintentionally—social normalization, nationalist aspirations, capitalist imperatives, the list goes on. The point is made strongly here only because, as Leah Ceccarelli shows, science has a long history of being understood as a discipline "with a straight path toward truth" that lives independently of normalization and extracts the "twists and turns they [scientists] took along the way on the path toward agreement."[12] The contemporary neurosciences are from the very beginning informed by and guided by present social and economic conditions.[13] The neurosciences are entangled by cultural conceptions of bodies[14] and capable of reproducing contemporary categories of healthfulness and abnormality within the very development of experimental work.[15] Accordingly, neuroscience cannot remain divided from the challenges or complexities of social morays and political disturbances.

The mythology of the phoenix says that it lives only ever as a reanimation of a predecessor. The magnanimous image of the brain may well "rise like a phoenix," but as a phoenix, we should never forget that it has a past. As for the phoenix, we discover in the flying beast an old imperial symbol of a blood lust for terroristic revenge; the phoenix imagery carries a connection to rehabilitation and reincarnation, which offers hope of destiny and liberation in the face of horrific struggle, war, and death. The phoenix is also out of the old tradition of Greek oracles, a symbol of Truth completely inseparable from the false appearance/real world dichotomies of Plato and Socrates. The phoenix moreover participates in traditions of allegory and verse, which are common to European folk songs and themselves serve social purposes.[16] The phoenix is, through and through, nothing of the present nor of itself. Yet it rises.

The force of the point about a brain that can be an oracle able to discern truth from falsity is not lost on those studying the social effects of the neurosciences. As Judy Illes says, a neuroscience of everything and anything, regardless of rigor or appropriateness to a case, will not be kept out of the courtrooms and halls of the legislator.[17] The explanatory power of, say, twenty-two men aged 18–29 living near Indiana University is more than enough to argue that violent video games increase aggressive behavior in the short term.[18] If personal experience or phenomenological knowledge is not enough to persuade, then neuroscience may well be enough. Neuroscience as confirmation of otherwise lived realities will embed within and serve to legitimize or delegitimize modes of living. Emerging neuro-technologies, developments in brain imaging, and neuro-surgical interventions will all continue to challenge ways of being and doing as well as raise ethical issues, including those around legalization and institutionalization. Issues of privacy, vulnerability, access, ability, etc. will all be subject to what neuroscience does with the brain.

Within cultural priority, neuroscience shifts (with us and through us) from particular brains to The Brain, thus moving from the concrete to the abstract, from the immanent to the transcendent, from the Many to the One. The Brain is a kind of Platonic Ideal. What is real? The Brain. How do we know? Look at those neurons; they just are; they are the given ("raw data"); they speak for themselves. Finally, the world of appearances is unmasked. We allow neuroscience to be caught up in materialism and idealism, realism and anti-realism all at the same time.

In transforming the brain into The Brain, we discover something to resolve a search for unity and meaning—the One, a transcendence that grounds us by explaining how we are and what we should do. The One, the enduring return, is the pursuit of ultimate, universal conclusiveness. But more basically, the One is a need to resolve inner inconclusiveness and solidify a shaky social field. The Brain rises like a phoenix, but the phoenix is nothing except an abstraction. What does it abstract? The ashes from which it rises.

Phenomenologically, Affective Realism is a yearning for a "Here You Are" for all mysterious dispositions. And Affective Realism delivers: it delineates in exacting technical detail sudden and intense feelings that shoot out like a geyser from who knows where. The idea is simple: the brain, that devil, cannot hide from us anymore. If there ever was a Freudian subconscious, then it, too, sits now vulnerable. The technical apparatus can and will delineate our secrets. Affective Realism promises our intimacies. We visit the museum to see ourselves through and through. And the visit is supposed to be satisfying?

We become the bowl of popcorn. The beer overflowing with Self-love.

But Affective Realism is inherently self-deception. Something stands truly against ourselves in the compulsion to make the visual va-va-voom of gorgeous neuronal complexes an exposure of hidden feelings. The giveaway that blows the whistle is the image itself. The materiality is so damn specific. The brain. My experience pictured in X way. The stroke of paint. The metallic lacing. The museum space. Narrowed. Sanitized. It's just not bloody or messy enough. Why would we be convinced this is a brain at all, much less a Self?

Standing before Chuck Close's *April*, we might come to believe that the girl hopes for something more, but she doesn't have a clue how to get it. And the closer we get to her, the more we feel the same way. That is the wonder of the work. In some kind of magical conjoining of form with interpretive promise, we realize that the closer we get to April, the more distant she grows, the more pixelated she becomes. The image transforms until she is nothing but dots and colors. She dissipates in pattern. The illusion is gone. The painter openly admits how it is all done once we follow her call to get closer and closer to her. The same must be true if we take a step forward and look at brain artworks. But Affective Realism holds out an arm, keeps us looking at the thing from a distance to avoid noticing that the construction.

Another way to think about Affective Realism, again as a thematic, is to compare it to the myth of Sisyphus. Affective Realism's endgame is the top of the hill with the big rock. The myth of Sisyphus, as a paradigmatic portrait of life as "repetitious, meaningless, pointless toil that adds up to nothing in the end," needs none of the usual "solace in producing and raising children," which just displaces the struggle and is "more of the same: adding zeros to zeros"; Affective Realism races right past any struggle, seeking after the glimmering flush of revelatory human constitutions.[19] Or Affective Realism finds perhaps another solace: the brain scan convinces us that we are getting somewhere transformative. In overcoming the Sisyphean task, humans would finally achieve power over themselves and others. Any hardship or repetitious toil may well be zappable or extractable. It's all brain matter.

Comparing Affective Realism to the myth of Sisyphus offers the chance to think in relation to Nietzsche's non-Sisyphean "overman."[20] For Nietzsche, the overman feels deeply empowered in the embrace of emptiness at the core of all human activity but transforms in living out that condition, thus overcoming the suffering of relentless defeat. That is Nietzsche's idea anyway. Affective Realism proposes that human technical activity, absent of any profound interpersonal reflection, eventually leads to a glorious future of self-chosen self-development. Here, it seems to me, Nietzsche's overman stands in opposition to the tendencies of Affective Realism. But the two could well be put into communion.

Let's imagine them in communion first: personal transformation through confronting an empty ontological core could be initiated from staring into the brain and finding only neuronal connections that shift and jitter. Indeed, psychologically overcoming the sense of one's own emptiness of Being might then lead to investments in the biomedical sciences that allow Affective Realism to shine. Yet, we can also imagine the opposite: if we understand ourselves through our brains and propose a "real" biological knowledge of underlying Self mechanisms, then we seek another solidity or transcendence. In this, we contravene Nietzsche. We make a promise to ourselves for a kind of freedom from our bodies through scientific means, and we enter into a non-acceptance of our state, which is precisely what Nietzsche hopes we wil deal with and dwell upon. Indeed, we do not need to become an "overman," really, if the old body can lead us triumphantly to a new one. I am rather tempted to say that Affective Realism, in being tied to the scientific real, sides with this latter image.

Freedom from the Sisyphean task first requires achieving a knowledge of the passions that drive us to push any rock over and over again. The brain acts as the material to re-shape the dispositions, those unavoidable, inculcated passions; this is Affective Realism. The brain becomes wet clay for sculpting. With neuroscience, we can see that we can be the artists of ourselves and so we seek the tools. All we need to do is grasp the form and know what the clay can do. We must master neurobiological materiality according to the theme.

Affective Realism, then, is much more than a "neuro-realism" or any straight-forward legitimization of what we already know.[21] It is a project to end the rupture. It asks us to use representations to settle volcanic dislocation of the Self, once and for all. In the glorification of neural networks, it champions the universalizable neuro-structure, which is presented as beautiful both because it is useful and because the images present visual structural completeness and interior symmetry credited to the Natural order, to human consciousness, or to human potential, or to any combination of all three. Neuroscience, take my breath away.

If the neurosensuality explored in the previous chapter is an emotional transform-ation of neuroscience into a tool for personal care and comfort, then Affective Realism is a techno-scientific tactic for making aesthetic sense out of the radical incompleteness and alien(ness) of our whibbly bodies. It wrestles with the recalcitrant and disorderly (and hence, the strange or the sudden) through the powers of neuroscience. It cracks a magnetic resonance whip to tame a lion—what Salter calls "unruly experience" or the condition of unassailable complexity amid constant change in material Being.[22] Putting a cork in the overflowing bottle of existential doubt, sudden and uncon-trollable feeling above all else, gives the Affective Real its functional psychological character. In cases where I feel suddenly disgusted, terribly scared, unreasonably shy, mesmerized or awfully sick to the stomach, Affective Realism attempts to settle an answer as to why it is I am this way or that through the neuroscientific brain.

But brain art cannot be blamed for playing in the pool of the Affective Real. Questions of strange bodily productions, the oddities of perception, seem, on the face of things, rightly delegated to the neurosciences. Just as much to the arts. Both sustain an ironical unknown in so far as neither can explain how knowing anything is possible in the first place. Brain art, viewed from within a tension between kinds of knowing (and the persistent unknown) in two different disciplines engaging the body, makes perfect sense.

Indeed, "fuller examinations" of the philosophical and humanistic problems of the body require techno-scientific responses when the body hides from itself, i.e. when reason and phenomenological reflection cannot of themselves come up with answers.[23] Turning then to machinic images of neurobiology only seems sensible, especially given that art always turns to technics anyway, usually engages the contemporary, and expounds on the ethical limits of modern exercises. Fur-ther, philosophical abstractions made in the arts seem to perpetually need more evidence to scrape out some legitimation. Measuring brains and making data part of the art process proves useful toward this end.

As philosopher Roberto Simanowski says, we are all today desperately in love with data, even if it proves to be a bad lover who offers us little in return and watches everything that we do like an obsessive boyfriend. Applied to the discussion at hand, making unities of art and neuroscience, at this present moment, allows data love to perfume the museum space. The artist can take advantage of the mathemat-ical exigence by performing elegance as inherent to the brain, even if it requires two tons of machinery, a month of algorithmic sorting, and a set of participants who won't disrupt the making of the imagery.[24]

A caution is worth spelling out. Tactical investigations of art-neuroscience projects should not be immediately derided; it might seem that I am speaking in this way. But the potentials, as should be obvious by now, are multiple. However, the idea that we need to circumscribe our remnant mystery from within the images that the brain sciences offer might, at least at times, overlook the fun in creative ambiguities and psychedelic imaginaries of art itself. Depicting brains is a captivating choice, of course, precisely because neuroscience is the field most directly, or most authoritatively, involved in unravelling the mystery of minds; but for me, what makes brain artworks truly worth considering is how they do not depict minds nor have the authority to determine brains. Their excitement and value is how they sit unevenly on the balance beam of seeing and unseeing, mystery and revelation; how they show the momentary even while tempting us to see the constant; how they investigate the singular even while tempting us to extend out what we think we discover there to the many. This is the crazy joy of them, even if many still do make grand pronouncements or depict brains with such conviction and seriousness.

As I will show in this chapter, not every brain artwork gives the viewer any clue that it is, itself, aware of the irony that it opens up doubts about material realities in its own engagement with scientific realism. Works doused in the Chanel of Affective Realism fill the room and are easy to sniff out, but they also leave a little sweat on the canvas, exposing the dirt of the workmanship. They seduce a little, sure, but in winking at us too long, they turn us off and undo their own directionality. That is, brain artworks can force the viewer to abandon any unequivocal claims about the supposedly life-altering power of neuroscience that they might themselves be selling. When viewing the brain artworks in this chapter, we can stand there slack-jawed at what neuroscience can do today to show us the complexity of our brains, and then we can simultaneously raise an eyebrow and wonder anew about the boundlessness of our bodies.

That is the fun of brain art—watching the artworks suggest meanings far beyond what the artist intended. We might herein imagine ourselves at a saucy cocktail party where a magnetic (resonance) personality is making the rounds, whispering unspeakable seductions. We can listen quietly, laugh at the spectacle, or sneak off to the bedroom.

Illuminating the mind: *Brainbow Hippocampus*

Cultural fascination with all things "neuro" coupled with the academic drive for innovative, interdisciplinary entanglements creates ripe conditions for collaborations between artists and neuroscientists. Multiple projects have garnered attention in recent years, but none as prominently as Greg Dunn and Brian Edwards' *Mind Illuminated* exhibit at the Mütter Museum in Philadelphia.[25] Displaying massive 22K gilded microetchings of neuron networks shimmering with gold, they dazzle audiences with scale and light, showcasing the visual magnificence of neurobiology with opulent effect. In Dunn's words, the microetchings "remind us that the most marvelous machine in the known universe is at the core of our being."[26]

The artists' presentation alongside a title that conflates brain with "Mind" unsettles, leaving the viewer unsure about the role of the body in the production of mind and the difference between mind, brain, and body. The sparkling neuronal patterns seem easily interpreted (or misinterpreted) as something like neurobiological personality profiles. As a result, one immediate reading of *Mind Illuminated* foregrounds an underlying humanistic impulse that idealizes the body's architecture and situates neuroscientific representations as exposés of "our being." Although this general neuro-centricity may well be a recurring theme in brain art, one piece in the exhibit, in particular, offers insight into the multiple meanings and relationships possible when blending neuroscience with hyper-technical approaches to art.

Brainbow Hippocampus is a colorful web of wild neuronal lines. The work is grand and gilded, cropped tightly so that only a glistening brain-on-display fills the visual space. Seen against the white walls of an art museum, the piece looks like God went crazy with a box of crayons. We are not sure if it's an electrical storm or a time-lapse video of Manhattan. Maybe it is an abstract painting done by a three-year-old with too many emotions. We're not sure if its flat or if its deep; the kid is either a genius or a monster. We might see a nebula. We might see a killer dissected for insight into the black-hearted beast.

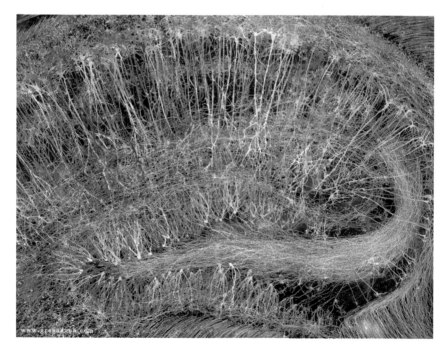

FIGURE 8 Gregg Dunn, *Brainbow Hippocampus*, 22K gilded microetching, 24X32, by permission.

As art intending a realist depiction of the brain, the work creates a repressed return to the impulses guiding old realisms—repressed because the twisting depth of the brain aims for adherence to the Thing yet self-denies its own access by highlighting the process of etching a metallic surface. The focus on the making of the piece, on the materiality that supports the performance, seems to undo the objectification of neuroscience. The idea that a scientific image exists free of human meddling and instruments or can be certified as "speaking for itself" dissipates when viewing this light show.[27] In brief, *Brainbow Hippocampus* connotes and denotes; it makes evident its practices even while it purports to show the actual neuronal pathways captured in an fMRI machine.

The work celebrates technical ambition, both of the artist and the neuroscientist; what is the difference, after all, the work seems to say? The piece may well hang in the museum as a witness to neuroscience's power to divulge human secrets, but as flashy techno-object, it tempts the viewer to closely examine the secret behind its "brainbow" decadence. Thus, the work does not, or cannot, achieve an uninterrupted look at the brain nor consummate a full repression of its constructedness. We cannot unsee its artsy technicity no matter how much the artist conflates brain with mind in presenting it to us. We might go so far as to say that the work inadvertently performs the old joke, "Who's on first," delivering the Costello line as well as playing Abbott, calling attention to the two sides of brain scan images— objectivist and rhetorical. We look-*through* to see "Me" and *look-at* to see an etching. But the work does not seem to get the joke, and nobody is laughing. If there is a joke in there, then it is Abbott talking about capturing "the actual" complexity of the human brain and meaning the "real" of material presence, while Costello thinks that he is talking about the "real" of scientific representations and then the "real" of psychological experience. Costello says with a confused look, "Who's on first?"

Here we can interrogate our fascinations and consider whether the piece undercuts or instead sustains the "seductive allure" of brain imagery.[28] Perhaps, we are not sure. If we choose to be star-struck, then we do not even think to ask if we should feel seduced at all. But why would we ever be star-struck? The artist does not seek to contribute to human knowledge about the brain. Dunn does not ever clearly relate these works to the history of art. So why are we mesmerized? Again, we're not sure.

We consider why our brains go all mushy in the museum space, why we feel compelled to gape in awe before giant glowing masterworks that efficiently hypnotize yet offer little in terms of education and do little (or nothing) to advance the neurosciences. Maybe the answer is in the tangle. But where? We feel lost. Maybe that twinkly web is a critical self-commentary. We want to see the irony. The most satisfying nakedness is always in the revelation of a hidden secret, an opposition driving us at our core but never before noticed. This is the bodily satisfaction of brain art. We're still standing there, basking in the glow of *Brainbow Hippocampus* searching for it.

We shift to a previous thought. Art history. Maybe the answer to its ironical play is in an underlying association. Dunn argues, in at least one article publicizing his work, that he is inspired by Japanese sumi-e ink paintings. The article states that "his interpretation of the art involves blowing ink around a canvas, or using a syringe to place it on the frame."[29] The association can be viewed as disjointed or as making perfect sense. On the one hand, sumi-e seeks after "the real" as much as Dunn. Typical explanations of sumi-e relate it to Zen practices wherein "reality is expressed by reducing it to its pure, bare form."[30] The belief that there is such a form is itself strange, especially if being staged as the brain through the scientific frame. But sumi-e processes do not much mirror Dunn's, so there is some disconnect. Despite the fact that neurons do look a lot like tree branches, which is a typical subject matter of sumi-e, Dunn never intends to leave much room for spontaneity. Sumi-e requires spontaneity, and the practice intends for the edges of the ink to bleed into the paper, for the paper and the ink to become one in a fluid kind of way, and for the artist to channel the paper or seek to become one with the work of art. It is no surprise why scrupulously copying a brain scan of another person to capture "the real" detail would sit unevenly with this association. If *Brainbow Hippocampus* is at all a reflection of sumi-e, then it may be, at best, a kind of Western, scientized funhouse mirror conception of sumi-e.

We feel unsatisfied. So we might then try conjure another art history comparison. Frank Stella's color field rainbows seems a good place to start. Like Stella's works, *Brainbow Hippocampus* treats "the brush stroke [read: etched line] not as a personal trace, but as the atom or building block of a picture."[31] Likewise, the components of Dunn's work are objectified in so far as the lines are not so individualized as to harken back to the artist. On second thought, maybe they do, just a little. But each line stands on its own, and we cannot but see them as neurons. Each neuron is imagined as outside of the artist, a visual connection to a memory or smell or maybe the meow of the household cat. But in Dunn's work, we do not lose sight of the representational Thing; we see blocks of color combined but get so close that the colors get wild and confused.

Dunn's artwork nevertheless retains some similarity to Stella's color field art in so far as *Brainbow Hippocampus* "is a continuation of the old modernist ambition to dissolve art into nature, which is really a way of using the trope of 'nature' to reinvent the artist as something more than an ego."[32] However, for color field artists, this unity of human–nature or of art–nature was achieved in foregrounding the accidental stroke of color, the surprising play of colors outside of strategy and planning. *Brainbow Hippocampus* participates in the same general Modernist motive with respect to the naturalist tendency (and the rhetoric of making art a practice about exposing Nature), but Dunn uses backlighting to add a range of color reflections. In this way, it is a technocratic dedication to many fields of color, as opposed to "color field art" properly speaking. *Brainbow Hippocampus* seems then a production apparently in love with the affective influences of scientific realism over and above the affective intensities of color itself.

In thinking out this comparison, we should observe that *Brainbow Hippocampus* seeks no minimalism to achieve its effect even if it does, like color field art, reinvent reductionism. So despite some overlaps with other pieces in art history that also fill the frame from top to bottom with brilliant and alluring color, *Brainbow Hippocampus* and all the other pieces in the *Mind Illuminated* exhibit stand adamantly as themselves. From my point of view, they seem to be more divided from art history than serving as an intertextual commentary; they appear motivated by a dedication to celebrate the brain and the wonder of neuroscience's aesthetics.

And they do a good job. We are wowed. And we should be aware of the fact that we want to be when walking into an art exhibit. The immense effort of etching thousands of neuronal lines is quite impressive. But we hesitate, perhaps, because we recognize that the piece milks a fascination with "neuro-everything."[33] And if it induces awe—an expansion of minds—then it probably succeeds because it remains disconnected from any obvious critique of brain images and does little to encourage the viewer to disarticulate neuroscientific interpretations from psychological self-evaluations. That is to say, rumination on the governmentality of the neurosciences is not a pressing response to these works. *Brainbow Hippocampus* does not encourage viewers to reflect on attempts "to shape human conduct by calculated means" nor to think through how neuroscience might be used to support programs "educating desires, configuring habits and beliefs" to scaffold power with legitimacy so that it can operate with consent despite being a disciplined and historically specific form of power.[34] These works, rather, sell beauty and complexity as Self amazement and then push us to indulge in the pleasure.

Scan 9: the Brain sells

Neuroscience has earned its reputation. Thousands of articles on brain regions, neural systems, and emotional regulations now fill over 136 reputable neuroscience journals.[35] However, the popular rise of the brain sciences is not due to the efforts of scientists alone. The academic conversation about the brain has always tarried right alongside the massive brain-boosterism of Capitalist consumerism. "The Decade of the Brain" also saw "Baby Einstein." The 81.4 million-dollar National Institutes of Health "The BRAIN Initiative" garnered media attention[36] but did not attract one-tenth of what was showered on "Brain Games" for mobile phones, brain energy drinks, and brain tricks for expanding one's own financial fluidity. TV shows and software platforms proliferate.[37] The Brain Bullet, billed as a "performance technology," for example, promises that it "zaps your brain with powerful commands ... Helping you achieve almost anything you desire!"[38] NeuroNation, as another example, uses its "science" webpage to advertise how specially designed brain games help users to "improve concentration ... Become more successful."[39] Old and young alike have chased the seductive promise of increased agency and longevity.

A related dynamic happened across the university. "The Neuro-Turn" in the humanities and social sciences emerged partially due to connections that could be made with the neurosciences, which would contribute to field-specific invention; however, it expanded quickly because of the interdisciplinary legitimization endowed by the more financially and epistemologically salient fields in the sciences.[40] That is to imply that neuroscience was, at times, not really needed in much of the work by non-biological fields citing it. No sociologist ever doubted that drugs ignited people's pleasure circuits, for example, just as rhetoricians and psychologists never doubted that individuals suffering with insomnia would have difficulty making decisions. No one ever doubted that the use of eye contact would end up having a neurobiological correlation to areas of the brain associated with social interactions. Nevertheless, roping in a neuron or two heightened the exigence, offered prestige, and enabled new funding opportunities.[41]

Regardless, some interdisciplinary investigations did discover compelling avenues, and neuroscience is a diverse field area with many streams of investigation, which have proven variably applied. As a result, citing a finding from the affective neurosciences where women look into their own babies' faces to understand emotional processing is something quite different from citing a study that correlates mate preference to brain size.[42] In the humanities, the specific outcomes of any crossover with such studies remains dependent upon the function and extent of the application, even as any deployment will always sit open to criticism amongst field-specific audiences.

Whether or not Freud's theory about internal affective unities could be effectively bolstered with appeals to neurobiology remains up to those fields desiring to use Freud once again.[43] Likewise whether the experimental mirroring psychotherapies of the 1970s could be shown to be based upon an underlying automatic neurological compulsion to reflect back others, or not, would be up to those using mirroring behaviors to try to make therapeutic breakthroughs.[44] Of course, older theories, such as those about the indisputable genius of certain musical compositions, would almost certainly find some additional biological backing with neurobiological correlations to pleasure and control regions of the brain.[45] But taken together, these examples suggest that there is some kind of future for the neuro-humanities—but absolutely one where researchers need to pay close attention to making a good fit between conclusions acceptable in their own field areas and what the neurosciences can actually offer. Up to now, it seems fair to say that the quality of these collaborative endeavors has been uneven. While some cross-disciplinary projects have achieved theoretical breakthroughs, others have unsheathed new authority for tenuous ideas or, probably more often the case, added support to ideas already established in their respective domains.

Some applications of the neurosciences seem to function to enhance the sellability of a new neuroproduct. As one example, a handful of neuro-marketing professionals have charged big bucks to Hollywood moguls under the aegis of teaching which film trailer is the scariest, and they do this by putting audience members in an fMRI machine or by hooking them up to EEG, measuring electrical activity from the brain. Presumably, some of the test participants would not be able to report the internal stimulation of their own scary feelings or might feel tempted to lie in a personal interview asking about the scary moments. Thus, the neuro apparatus could be touted as revealing more while having the benefit of showing when film directors elicit basic evolutionary fear responses. However, it seems reasonable that those in charge of interpreting the brain scan results could also (inadvertently) skew results with their own individualized methods and perspectives. Their strategies for interpreting "scariness" within the experimental design, for instance, is going to make all the difference. It will either align with the test participant's own self-reported feelings or not; if they do align, then one would wonder why the brain scanner was ever needed, yet if they did not align, then one would doubt the results, since, of course, the participant would not be reporting fear but be producing data to the contrary. Questions about what is the case, exactly, regarding the nature of scariness, how it is determined, and who might be more prone to tell "the truth" of a scary feeling—the person viewing the film or the person paid to conduct the testing—can be brushed aside when neuroscience is held up as delivering unassailable, objective brain data telling a kind of greater truth than even the person having the subjective experience.[46]

Neuroscience in Hollywood is not an isolated case of making brain research profitable or of shuffling some findings into situations where it can benefit other professions. Neuro-legal consultants peddle the most mundane and preliminary neuroscience studies to, God bless them, trial lawyers desperate for using a wink and trick to ensure favorable juries.[47] Precisely how much the neuroscience is overextended in such uses is debatable, but understanding "brain basics" for lawyers is incontrovertibly rooted in generalizations about processes of evolution, consciousness, and confirmation biases.[48] Yet, even the more specific and serious findings like "mirror neurons" have been applied in a hundred different ways across numerous other field areas, as is well documented, despite most of those applications being quickly condemned and overturned by neuroscientists themselves.[49]

This sudden "neuro-everything"[50] has spurred scholars working in rhetoric and cultural theory to critique the emergence of brain-based explanations for new education curriculums[51] and to call attention to new forms of discourse about what's "normal" and not,[52] including who is criminal, insane, right-minded, etc.[53] In the past five years, books with titles like *Brain Culture*[54] and *Brain and Culture*[55] and *The Critical Neurosciences*[56] have started to address the uncritical acceptance of neuroscience. These works give valuable insights into how sociality fuses with neuroscience processes and products. However, what may still remain depressed in the discussion are the ways that the current neuro-craze relates to a much older history of staging the body as dumb flesh and seeing the environment as exterior to bodies, as opposed to enwrapped with them; the division of mind from body and environment informs contemporary efforts to mediate "the real" and to define what is real within the narrow scope of what an MRI, fMRI, or EEG can detect.

The aim to declare, once and for all, a resolute way to decide the parameters of things (like "scariness") and to confirm knowledge (like whether a movie trailer *really does* scare people or not) expresses an old tendency. The search for a (more) reliable means to guide human life. Neuroscience is another ripple outward of a heavy stone that has been dropped into the lake of our body's awareness of its strange and unstable existence. When we feel confused about whether we can trust ourselves, when we feel tribulations that impact our quality of life, and when we have little clue about what to do next, we need a wave to ride somewhere.

We can also see Dunn's work another way. We can imagine it functioning like a Vegas magic show—there to draw us in but not any more about "reality" than the poof of smoke crafted by the magician. Moving back and forth to see the light flicker across the neuronal lines situates audiences as self-hypnotizing Beings. The magician does very little, in fact. The audience already holds the built-in compulsion to convince themselves that their own secrets can be revealed, if only they sway back and forth and submit to the hypnotist. The artist, like the Vegas magician, is there to remind us of how easily we are tricked and how much we want to believe in magic.

Of course, unmentioned as of yet is a remarkable and commendable fact about *Brainbow Hippocampus*, which is not immediately evident and which adds to its artistic significance. Principally, the work displays a rodent's hippocampus. Thus, any correlation to the human Self would be a terrible and hilarious mistake. Or perhaps not. The title, *Mind Illuminated*, productively insinuates that human and animal both have a complex consciousness and are worthy of comparison. The inadvertent comparison to animals by museum visitors—who are going, we suppose, to perceive for themselves only a human brain in a show titled in this way—interrogates our ideas about the outsider status of animal cognition. What of the animals? Are we so much different, we ask? Look at the brain, the artist slyly suggests, and please, see yourself—but be sure to look again and see a rat.

The move to give animal materiality its rightful place in the museum has not been appreciated in any of the reviews that I have encountered. Indeed, if *Brainbow Hippocampus* props up seduction as a central tenet for brain art—or the kind of brain art that likes to mesmerize audiences with a big show—then animal brains are included as a kind of foil that can undercut the "genius" complexity of the human as a brain. Animals look similar to us and complex too, we cannot forget. Dunn reorders our rigid hierarchy without telling us; the descriptive museum card only reads: "22K gilded microetching in custom frame." The subtle move, conducted under the nose of those museum-goers conversing about the magnificent complexity of human brains, might even suggest that we should overturn our own celebrations. We should stop pointing to the fact, as Dunn himself says to some ironic effect here, that "we have, at the very core of our being, the most complex machine in the entire universe."[57] The "we" might be re-read as all kinds of animals within which the human is only one.

But the more obvious reading of *Brainbow Hippocampus* may be the human-centric one, i.e. the rodent brain is elided. Because Dunn works with neuroscientist Dr. Brian Edwards to produce the exhibit and foregrounds human brain functions, the intent seems resolutely "to elucidate the nature of consciousness" within an oversized map of colorful neural connections.[58] Seen this way, it is no mistake that the rodent brain is not clearly presented as such and nobody seems to talk about it. The work falls back toward old human-centric ambitions. Indeed, by making scientific product and achievement the focus, the viewer imagines Dunn as a contemporary Leonardo Da Vinci seeking "universal, all-encompassing knowledge" in composing unforeseen beauty by deepening the scientific gaze.[59] For Dunn, like Leonardo, if art and science are perpetually enmeshed, then technical artworks follow from (or stoke) the old idea that human ingenuity and perception can combine to discover Truth and beauty in Nature.[60] Looking ever more closely, Dunn implies, eventually will reveal everything. It is all right there in the crisscrossed lines.

Despite crafting a "David" statue of the brain, such images inherently reject the body. *Brainbow Hippocampus* seems to unavoidably champion the brain-in-the vat. *Mind Illuminated* gives no glimpse of the body. The hyper-focus on a massive neuronal interior ignores the pumping blood and undulating guts, the tendons of the fingers, and the bones of the feet. The "Brainbow" recycles Descartian dualism as a beautific vision, ignoring the outdated nature of the conception. Yet there it is. Alive, seductive as ever. The Brain. As Descartes says, "I think, therefore I am."[61] Descartes, like Leonardo, might be proud of these images, but Leonardo likely would not recognize the absence of the body as the correct means to send a message of human empowerment.

Then again, the work might be self-effacing. We can turn ourselves back around when looking at it. The art critic could have it all wrong. The viewer is meant to compose a realization that internal views are insufficient for an explanation of cognition that is dependent upon immeasurable environmental suasions. In being hyper-zoomed and microscopic, the image must compel the viewer to imagine how much has gone missing. The work, then, functions like a cropped photograph of a family where half of one arm of some person remains, telling the viewer that someone took it upon themselves to have a go with the scissors. No, the title of exhibit suggests differently, we think: this seems, rather, to be a reification of cognitivism and the computational neurosciences, not an ironical self-play announcing the need for embodied and extended cognition.

The absence of the body is obvious and should be considered as itself a critical move; however, the neuro-centric framing of the exhibit derives from Dunn himself and, thus, probably is not meant to be snarky even if it is unavoidable. Dunn consistently stresses the idea that "brains govern everything."[62] For him, the exhibit has a defined purpose—to use the "unique power of art" to inspire audiences to learn about the brain and "to provide a piece of art aimed at professional neuroscientists that is as close as possible to complete anatomical and functional correctness."[63] The exhibit, accordingly, takes itself and the ability to access the material real quite seriously. The intensity of the algorithmic and networked thinking evident in the production and visual form of the work also drive home the point. *Brainbow Hippocampus* leaves the impression that the computational discourses dominating neuroscience are correct after all. We are, when looking at the etched lines that catch the light just right, a tangle of neurons not much different than the Internet. The immense structures of the Internet, we might even start to believe, are an inverted material mirror of ourselves. The brain is a traffic jam. The city is a brain. No, no, we shake it off. It's a trick. The light, the flash, a rabbit out of a hat. Vegas magic.

Mind Illuminated, as an "illumination," invites some critique for the promise that it presumes in its title while yet staging such a narrowed lens; but the exhibit—and works like *Brainbow Hippocampus*—also allow us to see the many complex readings that brain art can produce. If we walk away with merely one impression, or take just one side, then we miss the fun and overlook the dynamics of the show.

In highlighting *Mind Illuminated* as a complicated art-science offering, the exhibit might be able to say something about the power of machines, or of materiality more generally, to guide us and to persuade us, i.e. to be folded into our own rationalities. In *Mind Illuminated*, materiality seems, inside of the frame anyhow, to speak for itself, over and above the suasions of human discourse about it; this is at least partially what the academic turn toward "New Materialisms" intends.[64] *Brainbow Hippocampus*, we imagine, is meant to give access to rote materiality and thus communicate more to us (or more directly or intensely) about the power of brains than even a lecture from a neuroscientist could do. Indeed, one reason to think New Materialisms together with *Brainbow Hippocampus* is, as Connolly explains, that a New Materialism, like Dunn's artwork, emphasizes "systems and things" and holds a tendency to focus on "manifold entanglements."[65] New Materialisms also, like Dunn's work, propose a return to "classical pragmatism" hoping to combine "experiments and speculations" to prod the world's self-organizations. Thus, New Materialisms utilize "techno-artistic tactics" to stretch everyday "notions of explanation, system, agency, objects, morality, cognition."[66] In seeking to close the historical gap between "language, consciousness, subjectivity, agency, mind" and "biological material or the inertia of physical stuff," New Materialisms tackle a complex task, one that positions Dunn's work at the cutting edge of this intellectual trajectory. However, New Materialisms also risk cultivating ground for new places for objectifications and for inflated claims about the human by pointing to techno-scientific knowledges and a material outside that in its designation as such frees it a little from the usual social critiques of reductionism.[67] We can make the same critique of Dunn's exhibit. Art–science collaborations may feel new to us, but like New Materialisms, they might also risk, at moments, slipping into the ideologically very old.

Like many of the emerging realisms, New Materialisms might stew in seriousness sometimes about its dedication to Things and to ontologies within its institutional buzz, just as *Brainbow Hippocampus* appears so crisp and unyielding. In some ways, *Brainbow Hippocampus* condenses down a return to realism; it might be useful for distilling the flavor for us—like a beefy vegetable soup that must be cooked just right and just long enough, or else it will be too watery or stink like hell. That is, Dunn's work can expose the difficulty of the recipe. Moving beyond the constraints of post-structuralism in order to embrace the "real world" of material stuff risks privileging "hard" ingredients and not using enough picante pepper. Where is all the kidding around with neuro-intensity? Where's the big joke about how Ralph always said he was a good swimmer when he falls drunk into the party punch bowl?

Janet Wolff bemoans the general displacement of "critical theories of culture—sociological, hermeneutic, semiotic, interpretive" in favor of the "agency of objects" and an "embodied nature" such that the "power of images," she suggests, now rests in their "engagement of the material world."[68] There is, in other words, an outside that all images must seemingly give us access to, and *Mind Illuminated* certainly purports to give it. So why are we doubting its visual disclosure? Maybe because we have a feeling that *Brainbow Hippocampus'* visual complexity does not so much expose our own stupidity by being too dense and ornate to decipher but imposes a question about that complexity's immense simplicity because it offers an image that has been reduced by data structures in controlled environments, divided from worlds around it. If it dazzles audiences with a story of materiality's dominance, then it would probably benefit from a naughty heckler mocking the need to make representations that move beyond representation.[69] This seems to be what Wolff is saying, in so many words. Only a jokester takes a half-second snapshot of a brain yet claims to emphasize "manifold entanglements";[70] only a trickster chooses materials for a persuasive effect while also purporting to "give objects their due."[71] The true heckler in the museum might well be Dunn.

Taking narrowed material investigations as correlations to embodied experiences and manifestations of human life-worlds without any ontological and epistemological boozing represses the social exposé that Tony Clifton (or Andy Kaufman acting as Tony) performs when he stumbles onto the TV stage drunk and forces the formal daytime host to sing along karaoke style, tripping up the producer's schedule, embarrassing everyone, and exposing the staged fraud of the television schematic.[72] Where is the brain artwork deeply unconcerned with its own narrative of the real or its own internal coherence and stability? Where is brain artist totally flabbergasted by an effort to extend a moment in laboratory time all the way out to universal sets of personality profiles? Where is the ontological jokester in interdisciplinary art making?

Scan 10: the brain is a social technê

Appeals to governing biology transform the brain into a technê. Constructing neuronal matter as more primary and more agentive—and even more agentive than the influences of the social context—positions the neurosciences as able to redefine, and indeed remake, communal orders. Clues to the brain acting as a technê—a pragmatic social exercising of The Brain, which is a larger than life, abstract conception of what the brain can do—is more easily seen perhaps in instances where neuroscience is imported to restore or remake social stability.

Shipping off the brain of a mass killer to a neuroscience lab, as was done with Stephen Paddock's brain after the horrific Las Vegas shootings,[73] sends a signal about the social role of The Brain in relation to psychological destabilization and human motives. The biological organ of Paddock's brain did not deliver—nothing abnormal was found—but The Brain, as a social abstraction, would insist that the organ contains a viable explanation for the inexplicable horror. The path to ensuring no future instances of mass death striking out of the blue—an action in this case seemingly devoid of defined ideological reasoning or long simmering feelings of injustice from Paddock—necessitated the examination of the brain organ. Paddock must have had a tumor or a brain disorder, so many believed.[74] However, Paddock's brain showed nothing out of the ordinary for a man his age. Presuppositions that a range of "normality" means necessarily that some anomaly should be detectable in the event of any psychopathic behavior fell flat. As a result, The Brain as conceptual model, one governing Paddock's behavior, did not deliver on its promise. The failure in this particular case demonstrates at least two points.

First, any presumption that Paddock's brain would resolve the investigation exposes how presumed truths about The Brain have real impact on the world outside of the neurosciences. Popular ideas are used beyond any neurobiological realities noted inside of the lab. Neuroscientists may not be able to always predict how their findings will be used. Likewise, they cannot always dampen the expectations that a finding instills in people once adopting The Brain, nor can neuroscientists well resist the applications composed by those who work in courts and offices. We can expect The Brain to be applied to the most difficult and extravagant social cases.

Second, the functioning of the brain organ does not always align with what people imagine that the brain can do (in them, for them, and for society) in The Brain concept. In fact, from the perspective of The Brain, the study of the brain must improve to meet expectations. Conversely, the popular image of The Brain could adapt. In the example of Paddock's brain, The Brain could not fulfil its call to duty. But there would be a few ways to solve the problem and still keep explanations of Paddock's behavior inside of the narrative about the degradation of the organ, one which would, of course, side-step the influence of social and cultural troubles.

The first would be to argue that neuroscience professionals need more exacting tools to see the organ and to explain Paddock properly, i.e. The Brain gives no ground in the narrative. Another option, of course, is to argue that Paddock would need to be alive for neuroscientists to understand his brain's functioning, and the dead organ just cannot give up his secrets; so this is a matter of timing, and neuroscience does indeed have the capacity. Here again, The Brain narrative can be maintained. A third option would be to argue that neuroscience holds insufficient conceptual models in this first place and cannot currently understand how brains interact with bodily chemicals or cannot comprehend how brains exactly compel behavior. In this instance, there is a theoretical problem, and The Brain diminishes. A fourth view, more radical, would dislodge The Brain entirely. This view would look outside the brain to find an answer and seek after a collective. This would mean composing greater complexity, turning to an entirely different concept of a brain-body-world, one foreign to The Brain but perhaps better suited to explain such cases.

Despite these possible alterations, the failure of Paddock's brain to align with The Brain's social promises takes nothing away from the fact that The Brain, in its current conception, may well continue to fill similar explanatory roles in the future. The power of The Brain as human distillation, as creator and governor, is not diminished by a single failure, despite how public the case was. The interjection of the neuroscientist into the courtroom to help resolve a question of sanity for a killer's defense or to see whether a defendant retains a memory or not offers a clue to future working tasks for neuroscience. The Brain will be imported, added onto valuable neuroscience, to convince juries and resolve institutional faults and uncertainties.[75] Watching the government develop health and education policy based on neuroscience research raises a question of whether the force of The Brain is really what is being put to use.[76] In like manner, watching security agencies turn to brain scanning technologies in order to decide who to interrogate at the airport exposes the extent to which neuroscience can and will function socially, driven more fundamentally by a popularized concept of The Brain, which is a dedication to stability and control and, thus, to decoding and regulation.[77]

The brain being scanned inside of laboratories may not always entail the specter of The Brain, but The Brain integrates neuroscience with our desires. And more practically, The Brain becomes an appeal to achieve tasks, often ones not originally intended by neuroscience. Sometimes this "achieving" is creative application of the metaphoric force of a brain finding, such as the way that the mirror metaphor in "mirror neurons" inspired new forms of psychotherapy and alternate ways of working with children.[78] Other times, this "achieving" is the successful assignment of overwhelming agency, such as when "adolescent brains" are declared to be unavoidably susceptible to rewards and risky behavior, regardless of an individual's personality, upbringing, culture, or historical exposure to types of behaviors.[79] In that case, the concept of the adolescent as a brain establishes grounds for new family therapies, government oversights, and child education policies. Whatever way The Brain "works," it will continue and maneauver through legitimate observations of the brain organ. And all along the way, The Brain will grow a network of agentive actors.

Dying for the brain: *Neuro Memento Mori*

Memento mori are artistic representations, often of human skulls or of heads with the skin partially peeled away. Memento mori depict rotting flesh or evil-looking snakes swimming devilishly through eye sockets. Sometimes memento mori contain slithery ravenous snails tearing into meaty cheeks. Or they show the scales of time weighted to one side. Maybe a burnt down candle dripping wax. Memento mori reveal human material fragility. They remind us that we must die; but more so, the artworks "must in some way make us want to change our values."[80] These macabre pictures, according to Garber, ask us visualize "the nature of life and the limits of mortality," as when Shakespeare has Hamlet speak to a skull and say "remember me," which is "a reminder to Hamlet of his own death as well as that of his father."[81] Fitting to Hamlet's obsession with finding the killer, the skull in that scene, once conceptualized as a memento mori, launches a psychological shift in him and ushers in the sharp awareness of what he must now do. He sets off on a new journey.

Intrigued by the emotional intensity of memento mori, visual artist Jane Prophet collaborates with neuroscientists at Aarhus University to learn whether brain scan imagery might detect the material correlates of revelations about human mortality during confrontations with such artworks. She wonders whether a religiously inspired meditation session focusing on the inevitability of death will prove equally impactful as staring at a memento mori artwork; accordingly, she explores if the emotional effect of either experience (meditation or viewing the memento mori) will be visible in brain scans.[82] In this way, Prophet not only sets out to measure the allure of death (or of thinking about it), but to discover the source or limit of memento mori's psychological power. Her starting point is her own meditations on memento mori images while undergoing brain scanning. She thus personally interrogates the impact of the art historical phenomena and pursues an answer partially through her own self-reflection. By adopting the role of the test subject, she interrogates the capacities of functional magnetic resonance imaging (fMRI) to be able to provide an answer to what, exactly, drums up such feelings while also looking for a relationship between her feelings about the task and the process of being scanned.

The art-experiment proceeds in two rounds. First, Prophet lies inside the brain scanner and views the creepy, haunting images of memento mori amid other control images of living people. In a second round of scanning, she focuses on her own death through Buddhist meditation, imagining a situation where she dies right at that moment, feet growing cold, loved ones burying her, all the trauma of passing away. She records her experience with a video camera. The audience watches, intrigued. Viewers seemingly will ask with baited breath: what *is* that emotional experience of impending doom that we repress? What happens to us, *exactly*, when we start to imagine that we must die?

Starting this way, with these questions, gives neuroscience a chance to shed light on Prophet's own phenomenological accounts using different means. But in this Affective Realism—or shall we call it sheer optimism?—she stages the possibility of a reversal. The artwork first invests in charting what she does not understand about herself and, thus, presents neuroscience with the opportunity to offer a fresh, unmatched narrative of the machinations of the mind through examining brain functions; but her work then, second, questions the readiness of the field to offer such a narrative. The Affective Realism embedded in the impulse to create this artwork has the capacity to turn out to be but an illusion.

Prophet's starting point pretends to allow neuroscience the agency to decide the affecting power of memento mori, which is, lest we forget, long reported in art history, sociology, philosophy, rhetoric, etc. to be downright affecting. The audience wonders why we should doubt so many scholars? What more could we learn? Besides, do we not feel it ourselves? In fact, if memento mori do what they do and "make us want to change our values,"[83] then asking for neuroscience to find a biological mechanism seems off-point. The value of memento mori shifts from inner transformation of lived experience to a stimulus for tracing neurobiological triggers. So Prophet stands, in some way, opposed to the memento mori and to the bodies of her audiences as she aims to delineate for them, with a brain scan, how much, exactly, these artworks might affect them.

Here, we also think of the Buddhist monks who developed the meditation practices from which Prophet draws. The proposal motivating Prophet's artwork risks appearing antagonistic when asking whether their meditations on death can compete with art, or even that they should be measured or studied in this way. The whole question of the art–science collaboration, once seen as staging ancient philosophical and phenomenological knowledge against the new "realities" of the neurosciences, might seem quite silly or outrageous. We wonder if Prophet is taking the piss. But out of who? Neuroscience? Buddhist monks? Us? Herself? Of course, a major university is co-sponsoring the project, and we know Prophet's skill from her previous work; so we press forward. But we do so with reservation. Or, perhaps more strongly, we start to think that Prophet not only challenges neuroscience to prove itself against age-old practices but that maybe she also questions the value of humanities scholarship.

The transfer of the question about the *actual cause and actual impact* of human experience from the humanities to the neurosciences presumes some missing piece in the humanities. Neurobiological function may, who knows, add a crucial insight to the puzzle. Something about serotonin or decision-making areas or the thalamus, we presume. We're not sure what to do with that information, but as humanists and art lovers, we remain open to the possibility that it will radically alter our views of the body and mortality in unforeseen ways. Prophet seems to suggest the exploration is at least worth trying. So, again, we trust her spirit and follow her into the woods.

But the strong link that Prophet implies between the neurobiological and the phenomenological is also a way I presume for her, as an artist, to stage a social situation that is, itself, worthy of exploration. The broader purview of the work does not escape the sociologists or rhetoricians in the room. Asking for neuroscience to handle the question of the human with technical tools and bio data—or handing the question right over to them with a sly smile—challenges the priority of the neurosciences; it is a task test. Seeing this as a structured reflection on neuroscience being prioritized as hotly explanatory of a social arena (we are reminded of headlines like "Meditation really does relax you!") positions the audience to instigate critical reflection.

Ultimately, the neuroscience results are reported, and the audience deflates. In Prophet's documentation video, a neuroscientist tells the viewer that her brain scans show very little difference between the viewing of memento mori and her meditations on death. The conclusion, according to one neuroscientist, is that "the disgust area of the brain" or "the body recognition area of the brain was lighting up" while Prophet mulled her own death but not when in the control state.[84] Little more conclusion is given. The audience simply does not know what to do with that information. We wonder if Prophet disgusts herself when thinking of death. Is death disgusting to us too? We mull the remnant possibilities, and we turn then to her reflective self-reports.

To consider the work now, we must go back to its starting point: Prophet wants to know whether neurobiological events—and seeing what neurons might be doing—make any felt difference to her own emotional or spiritual life. She is, of course, interested in what the neuroscientists see when examining her brain; but she also wants to understand if she herself changes during the experiment. This is likely the reason that the laboratory results feel so patently unsatisfying; they are terribly mundane compared against Prophet's rich descriptions. Watching her expressive video, we think that Prophet might be using neuroscience for her own gain, not to sell an impenetrable Affective Realism, not to share the secret of human psychological repression, all told, but to make for herself an encounter with her own mortality.

Here, art process takes precedent over product. Even as the art–neuroscience collaboration stumbles over the recycled terminological wah-wah of neuroscience heard a thousand times in previous studies, i.e. "disgust areas light up," Prophet's interpersonal journey glows. In contrast to the depressingly poor neuro-conclusions, Prophet's documentation of meetings, her embodied sense of lying inside of the scanning machine, her meditations on death, the suspense that she shares on camera when not knowing the results, leaves the scientific description empty by comparison.[85] What stands out in the project is her—her reactions, her emotions, her vibrant inner life—not the jargon of neurological systems that sup-posedly do X or Y. What strikes the viewer is her affectation for the process, the rollercoaster of the experiment. The art is a process made internal to her emotional well-being. And we can go along for this ride.

But what is the image of neuroscience here? The dullness and smallness of the conclusion in comparison to Prophet's own detailed discussion presents a flaccid neuroscience, one too stuck on rote categorizations of brain systems to say much at all about the vividness of mental life. Tellingly, at no point in Prophet's documentation does Prophet ever say that she felt "disgusted." This obvious disjunction with the neuroscientist's interpretation makes the brain scanning task (at least in her video) appear clinical but also pointless. When placed in contrast to the neuroscientist's brevity, Prophet's long elaboration of how she would miss her husband if she died and would miss "the experience of being in my body" because she loves "warm sunshine on her face" and would miss her friends—all narrated while inside of the scanner—highlight the sensuality of embodiment over and above any emotional connection to neuroscience.[86]

As a result, the immediate way to think about Prophet's art–science experiment is that it resists neuroscience as the One to describe the human. Indeed, the viewer might leave with the impression that neuroscience is pre-informed by social or embodied knowledges, i.e. the reasoning that we avoid thinking about death because it must be too disgusting does not sound very technical nor correct. There is a distinct air of folk knowledge in the interpretation. And the researchers are not made to be endearing whatsoever, given very little screen time in the documentation in contrast to how much we see and hear from Prophet.

Of course, Prophet's work does not end with the lab experiment. She is process oriented, certainly, so the work moves on. It also includes a digital projection of her fMRI brain scan onto a 3-D printed head, which has been modeled to be identical to her own. Presumably, she wants something to display. But the result is also that the audience then sees her face frozen in cold white plaster with the bright colorful lights of the fMRI added; this has effect. When rolling across the plaster, images of the brain jump back to life. In some sense, the multimedia sculpture reverses the emotional relations of the video. Neuroscience looks to be vivid once more, teeming with emotional energy, while Prophet dies in her death mask.

If the documentation video resists connecting neuroscience strongly to Prophet's individualized feelings, then *Neuro memento mori*—the 3-D printed artwork—gives some priority to neuroscience's technical capacity to make *more real* her emotional journey. At the same time, however, what we encounter is a projection over top of a plaster head as a colorless form, a human prototype or tabula rasa. The strategy distances Prophet's experience; the look-alike mannequin just cannot, perhaps ironically, compete with her flesh to show the viewer her Self. In this, the work retains a skepticism about the capacity of neuroimaging to produce valid one-to-one correlations, despite all appearances. Prophet, indeed, hints at this disjunction. She notes that her scan displays but a second's movement of oxygen across a tiny slice of her brain while she lies perfectly still in the scanner.[87] Thus, we are almost prompted to scoff at neuroscience while being asked to equate her feelings of death with the gorgeous visual wow of the animated scan. Splashing colorful blobs across a dull plaster head gestures to an inevitable breakdown of an ideology invested in producing amazing principles about her actual flesh, intelligible principles that can be applied directly back to what constitutes humanness.

However, if the emotional life of Prophet falls away from the work, then it is still possible that the scan projected across the molded head could lead viewers to contemplate death. The scan becomes a memento mori, of sorts, by insinuating that brains are the heart of the death matter. The work returns to the idea that neuroscience tells us how to speak to the living human condition and how to grapple with death. Within this return, *Neuro memento mori* creates a smart contrast: the work situates brain scans as able to stir up our most secret feelings about mortality while positioning the scanning machine as what is needed to comprehend but also battle mortality. We recognize that the printed output—the scan— affects us by forging for us a possible treatment amid a new medical regime. And we feel that we need something like this. We better look closely. But we're not sure what we are seeing in the flashing of brain scans. We leave a little confused, or confused perhaps that we are not disgusted by the experience but disappointed.

The priority given to brain scanning in *Neuro memento mori* can seem, at first blush, like a realist vindication. But it does not deliver satisfactorily on this, and probably intentionally. Thus, Prophet produces a work in my view that has more in common with the abstract artists of the 1940s and 1950s than with, say, the impulses driving Gustav Courbet. Prophet seems to reject fidelity to nature, asserting only superficial alignment—like surface overlay of a projected scan—between herself (or her meditations) and neuroscience productions. It really is a tease. It is probably going too far here to craft any strong relation between Prophet and the mid-century abstract painters; but there might be something there. Artists like Pollock splash and scratch because "of a breakdown of belief in the ability of man to order his experience in terms of general, intelligible principles which can be directly related to the external world itself."[88] Prophet's work, from this angle, achieves the work of Pollock by acting like Corbet. That is to say, the artwork asks viewers to consider the aesthetics of the complexity of biology and to engage its technical discourses while yet pushing past the prosaic celebration of brain science and the induced awe of "the real here to be unmasked."

In the non-scientific singularity of her experiment—a scan of one brain, and a highly "biased" art-informed brain experiment at that—Prophet presumes to transform that meager data set into an instructive colloquium for how to better understand our deepest feelings (about death!). We do not think her so audacious. We see the rhetorical move there. And it is made all the more forceful by the fact that she does achieve this faux-universalizing by exercising the wondrous technicity and authority of the neurosciences. Here, only with great pains could we ever align Prophet with realism, or an Affective Realism, in my terms. Indeed, we cannot. Prophet adopts little objectivity or scientism in offering up one brain and the hyper-individualized experience.

Something similar can be said about the *Neuro memento mori* sculpture. Despite the neuroscientific real being foregrounded, the brain scan on the plaster head is mostly a stage prop. It is ornamentation, perhaps, of our own imagined image of the prototypical human; cue the beautiful blue and red coloration. Sure, it could inadvertently position the mystery of human life as nothing but sets of neurons firing; however, the visual effect is one of raw materials peppered with painterly abstraction, hypnotizing disco pop, fun and games with the mirror and flash bulb. It's not as serious about disciplinary neuroscience as it appears.

Moreover, we can advance the thesis that Prophet aims not to seduce the viewer with Affective Realism because the viewer cannot help but see the artist and confront a self-portrait. As noted, the sculpture and documentation video seem mostly about Prophet and her unique experience. *Neuro memento mori* struggles to be turned outward and made into a work about the viewer's own reflections on death. The artwork is too confessional in tone; the fMRI scanner acts as a kind of reality TV show confession booth for Prophet. *Neuro memento mori* is not an interface for the audience to meet and greet death but for the audience to experience Prophet's inner divulgences.

Consequently, the artwork playfully draws attention to the Subject–Object dynamic of brains in neuroscientific narrations of Self. If the artwork is most pointedly about Prophet learning about herself by scanning her brain, then it situates the brain as the distressed patient working out fears of death in conversation with the same brain who plays the role also of the psychologist unwinding the complex realities scaffolding those fears. The brain is assessed and assessing. In this, the artwork is Object and Subject; thus, it is both comic and tragic. Neuroscience leads the patient on a journey to self-discovery, through the gates of trepidation, eventually to peace—but we cannot forget that the psychologist is the image of the brain that "speaks" to the neuroscientist and to Prophet while the patient is yet the living brain. Neuroscience in this work is truly self-medication. We laugh and cry.

Here, we are tempted to say that any realist impulses crash onto an existential beachhead. The human, as emotional Being, withers in the dreadful sight of a silent brain that could care less about the human. Even if the sight of the brain is packed with confident, warming colors, the scan sits there so damn indifferent and immovable. In feeling this, or in producing this feeling in me at least, Prophet successfully captures some central aspect of memento mori artworks, namely, *Neuro memento mori* represents a long psychological process of getting to the point where one is willing to denigrate the mind to rote and careless matter. This seems the central point of memento mori. The scan pushes the viewer toward an awareness of a slowly rotting body that lives yet is already living at a distance from the scan, and already changing, already dying.

Still, *Neuro memento mori* does not seem a good description. The artwork simply does more than perform a memento mori. Yet, at the same time, it does not meet all of the criteria: viewers reflect more on Prophet's self-feeling than on themselves when staring at it. But the project remains highly compelling, nevertheless, primarily as a manifestation of a tension between seeing neuroscience as able to add to the human experience yet say very little, as shockingly specular yet not seeing much, as playing inadvertently into discourses about a deeper truth to human nature while deploying formalized, often predictable, experiments to do so. *Neuro memento mori* may well begin as a form of Affective Realism but one that, quite satisfyingly, compels audiences to tumble down from the epistemological mountain into the valley of doubt where neuroscience has trouble explaining phenomenal experience. Of course, Prophet never voices too loudly any criticism of the interdisciplinary art–neuroscience exchange that allows her to make the work and feel so (Disgusted? Surprised? Sad? We're still not sure.) attuned to her own mortality, but the creative act does the talking for her.

Scan 11: the neuroscientific Self performs Self-extermination

If the object of the Self is the brain, then the object of the brain is neurons; the object of neurons is molecules; is that where it ends?

We obsess over interrelating these, and we have good reason. The memory, how it locks in place or suddenly lapses; the deepest love, how it thrives or intolerably dies; the ability to listen for long stretches or be distracted at every turn of phrase; these can be addressed, to some extent, by neuroscientific insights. Tackling such concerns aims to shorten the suffering of loss and restore what feels like rightful function, for ourselves as much as for our families and communities. Helping people retain memories for much longer or develop greater forms of compassion are undoubtedly important efforts, yet there is more to the story.

A psychological impulse might emerge when the brain and mind are set in correlation and sold as, basically, one and the same. There can be a kind of personal gratification—we should note for some more than others—in ripping open the mechanisms of the Self, in seeing the Self as mechanical and manipulatable, exposing personhood as a story we maintain to tell ourselves. The gratification is in exposing a fraud, a fraud purporting to have great substance; it is the unveiling of the poverty of one pretending to be rich—and the one doing the exposing is the rich one.

We should not presume that all of those who are brain obsessed are utilitarian or pragmatic only. As Žižek says, "true passion is not only 'hindered' but also even encouraged and sustained by the prospect of the 'gallows' — in other words, true 'passion' is uncannily close to the fulfilling of one's duty in spite of the external threat to it."[89] Žižek is herein arguing that Kant is resolutely wrong when "implicitly equating" external and internal Law by suggesting that existential threat keeps Law in place. Drawing from Lacan, Žižek takes a different angle. The snubbing of the Law, the act that violates, produces the pleasure, Žižek argues. Cheating on one's partner, Žižek says by way of an example, produces enjoyment precisely because of the threat of "the gallows."[90]

Those droning on about the nature of the Self—how the whole thing comes together in X and Y brain system, how neuroscience is the slow uncovering of the illusion that is you and me, how the brain must produce the totality of experience—swim in the lust of witnessing their own Self-destruction. Perhaps it is the equivalent of the inability to look away from the train wreck. It is the slicing open of the frog in fourth grade science class, feeling mesmerized by pinning back the skin, which makes your own skin tingle. That connection, skin to skin, is the violent and sick pleasure of the cut. To seek to understand one's own Self through describing the illu-sionary production of the Self in the brain is like telling the joke of contradic-tion with a deadpan straight face: "I'm terribly sorry, I don't speak English." There is an odd and often uncomely pleasure in testing the limits, in seeing how much we can stomach our own Self-elimination, the recursive explo-sion, the snake eating its own tail, the deconstruction of any account of Me.

The question is why. Why would this be pleasurable in the least?

To re-conceptualize "Me" as sets of neuronal connections—that is, as elec-trical and chemical exchanges—may well increase a feeling of commonality with other living things. But that immediately seems not to be where the pleasure comes from. Perhaps revealing something that was at one time too mysterious and too dark to ever have been breached, brings about a thick self-gratification, a hearty pat on the back—which would prove quite ironic given that the discovery producing the gratification is an illusion of neuronal forces, not anything of self-made ingenuity, and really nothing from which true "pride" can be born.

Another option is that thinking about the Self in a neuro-reductionist way pleases the investigator because it adds security through the technical advancements of the neurosciences, promises a better quality of life in the future, and, in this, the pleasure is a sense of increased agency outside of other forms of authority. With neuroscience's penetrating capacities, in other words, we can see a way to remake ourselves and break free from domineer-ing forces and oppressive histories. Maybe something like "go your own way" gives the pleasure?[91]

Yet the root of any pleasure derived from a Self fully "purged" of mystery or cracked open like a watermelon may yet (or also) be an inward protec-tionist response to the traumatic realization of the emptiness of a Self reduced down to neurons. The mind dispersed amongst mindless matter needs to see itself as superior to go on. Perhaps the pleasure results from feeling seriously through the nature of the discovery. Indeed, what else can people be expected to do with brain descriptions and discourses but look to find some pleasure in being nothing but a brain?

Here is what I'm driving at: can any human really be expected to replace a spiritual or metaphysical Self with mechanisms of rote biology dispassion-ately? I don't think so. Fear and pleasure are part of the game. They cannot be exercised through the sciences of the body. Reactions will range from anger to pleasure. And in fact, it may well be the case that neither anger nor pleasure about the state of the Self is warranted from what is agreed upon in neuroscience today. Overall, though, we must consider the role of feeling in biomedical assertions, and in particular, think carefully about the ways that interpretations regarding the meaning and impact of neuroscience findings influence the emotional life.

If Žižek argues that internal feeling does not actually keep external Law in place, then what he means, more precisely, is that pleasure does not keep external Law in place but, rather, leads, more likely, toward the violation of Law. Fear, more exactly, keeps the external Law in place. Pleasure is invested in destruction of Law. We might then try to argue that the neuroscientific inscription of the Self as a brain is held in place by some kind of fear—of being wrong, or of returning to non-scientific descriptions of the Self, or perhaps of being Lawless, or being set adrift.

We can follow the idea a little further: any "Law" of the brain, a Law that can dictate the accuracy of descriptions of the human, could also be challenged with pleasure, namely the pleasure of breaking that Law. Performing the pleasure of reducing the role of the brain in the production of Self or the pleasure of feeling through the whole body in an environmental co-being or of understanding the renewal of Self as a mystical reunion with spiritual energies enacts a rebellion against Law dictated by the brain (here, we could also say The Brain). Destroying the brain as ruler brings pleasure too. It is the pleasure of insurrection against technoscience and related conceptual configurations harkening back to The Enlightenment and Modernism. But in that case, the mysterious suasions of bacterial and viral life, the energy generated from within the depth of the ocean, or the unknown force crackling within the colossal expanse of the ungraspable universe full of billions of suns and innumerable galaxies would be precisely where we would find our freedom. How do you destroy that? You can't. So you revert to fear. But a healthier fear no doubt.

Willing the brain: *Interactive Brainlight*

Interactive Brainlight is a sculpture in the shape of a brain composed of acrylic perspex sheets inscribed with lines to resemble neural circuits. The work appears semi-translucent but is able to glow with various colors, depending on signals streaming in from an EEG headset worn by an audience member in the museum or gallery. As artist Laura Jade explains, "The installation is controlled with a wireless BCI (Brain Controlled Interface) Electroencephalography (EEG) headset that creates a dynamic light and sound experience of live interactive brain activity."[92] Detection of alpha waves activated by meditation make the sculpture blue, theta waves detecting focus change it to green, and beta waves formed from excitement make it red.[93]

In being able to consciously alter the color of the sculpture by calming one's mind or by thinking funny thoughts that excite brain waves, the work allows "a level of mind-control over our brain's bustling electrical activity while simultaneously witnessing an internal feedback loop created by our constantly changing perception of the artwork."[94] On a surface or somewhat trivial level, the functionality gears *Interactive Brainlight* for two purposes: (1) to learn the underlying "signals" that our brains produce under specific emotional conditions and to see if we can alter or control these signals; (2) to display our feeling states in public, to connect with others in shared situations where pleasure, anticipation, or exhilaration might be evoked, as when viewing the artwork in a museum. But techno-artistic play with mind control also points toward other more substantive functions in a neuroscience society; namely, the Self makes visible its affective motions to revise or redirect the body's affectivity and alter the Self, and others are allowed to participate in regulating or guiding the Self.

As a sculptural image of a brain reacting to human brain waves, the work creates a psychological and material loop. The Self is first interpellated, or "called forth" in Althusser's terminology, as a brain.[95] By controlling the brain, the Self thereby takes control of itself. Automatic affectivity is subjected to the domination of conscious thought about those resonate signals; the technical apparatus and attendant color scheme serve as a guide to one's own affective states and how to think about them (as calming or excited, etc.). This creates a feedback loop and gives affect additional definition, compelling specific emotional appraisals in well-known emotion terminology.[96] These appraisals then inform the process, which spurs more feelings and then more decisions about what to think and how to feel about what appears. In this, the technicity of *Interactive Brainlight* is a creative self-exploration, a way to revise one's own feelings in real time. It points toward a future where machines will help us to regulate our affect states and control bodily responses or where machines will supersede and anticipate our states for us. We can easily imagine machines showing us when we are getting too excited, too angry, too emotional, and perhaps they will work to calm us down again.

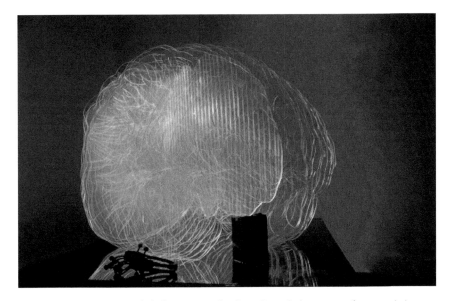

FIGURE 9 *Interactive Brainlight* by Laura Jade, from LauraJade.com.au, by permission.

The technicity imposes something like its own "will," from within its designed capacity. The participant may practice a form of self-mastery but only through and within the machine's affordances. Just as museum visitors imagine themselves in control—and, in fact, they do seem to expand their agency by finding new ways to experience their own experiences—so are they being controlled by the technical capacities and arrangements of the sculpture.

Put in critical theoretical terms, the work can be read as both a "Will to Power" and a "Will to Freedom."[97] With respect to the former: the participant energizes her Will to Power, learning how to exercise greater control over herself but also over the material environment, namely, the sculpture and the reactions of others in the room. In the performance, the participant's Will to Power acts as a seal or as a connection between the psychological (desire) and material (brain action) loop already implied by the sculpture, which also creates a recursive play amongst a Self and the Outside. Put another way, the promise of greater control over Self and Others is visible in the sculptural form of a glowing brain in a museum space and can be positioned as especially tantalizing because of the seamless connection to the Will to Power, as Nietzsche describes it. The sheer size of the brain sculpture seems to communicate what Nietzsche says about a Will to Power: "A living thing seeks above all to discharge its strength—life itself is Will to Power," or a will to overpower, as Walter Kaufman says when interpreting Nietzsche.[98] This is not merely a fight for recognition but a drive to be a convener of new realities within the action to keep on living; a Will adheres to a principal movement inside of all bodies to thrive in differentiation with other things, to survive in a physics of Wills.

Consequently, if the participant's Will exerts power over the sculpture, then it also acts in response to at least two external Wills, which spur the participant to sit there and generate a satisfying experience for herself and others. The first is the artist's. Laura Jade exerts Will not only in designing the techno-sculpture but also in encouraging others to see the experience as a playful engagement offering them revelations about neuroscience and the Self in an entertaining way. Thus, the structures of Capitalism—be they subconscious as well as materially locatable—also exercise a kind of Will, which manifests like an apparition of glossy media magazines mixing identity formation and gratification within the adoption of new technologies. The Capitalist imposition or Will, in other words, is the way that innovation does not escape discourses of self-control and self-improvement; the same could be said if the artwork is viewed as a kind of advertisement for the adoption of neuroscience technologies. *Interactive Brainlight* engages users with the force of these Wills but also allows them the ability to impose Will and to obtain gratification in self-manipulation while garnering praise from others, both for being vulnerable as the user and for successfully making the interior life into a technological stage show.

With respect to a Foucauldian "Will to Freedom"—a term deployed by Stephanie Batters to describe Foucault's effort to realize new structures of power through critique—the artwork functions in the same way that Batters wants critique to function. This requires some set-up. The basic idea is that critique requires "full immersion into the present" to think through the influences, which might lead then to a productive rumination where "discourses of power and truth" can be unmasked from within the experiences of the individual. In this mode of critique, the individual is then armed to try to escape any totalitarian social constructions and may produce new environments according to what feels more free, so situated.[99] In asking museum-goers to alter the colors of the sculpture and revise how they feel about it—to see if they can, and if they dare, resist excitement about the work when seeing others' mentalizations released to public scrutiny—*Interactive Brainlight* tests how well, and if, people challenge the work's underlying promise to generate new experiences. The sculpture, in some way, requests its participants to resist or interrogate the happy call to "gain control over the mind."[100] The act of putting on the EEG would be, in Batter's terms, a call to critique. The implied challenge of the work is to discover all of the brain states and then learn how the reactions of the audience members contribute to or inhibit the participant's goals of shaping brain states. A Will to Freedom, as a process of critique, might then take this as a mode of understanding contextual power relations to learn what is, in fact, being asked of people living in that region, i.e. how the functionality of the work and how people use it reveals what *should be* in the broader sociocultural arena and what is exerting Will. Consequently, the artwork can be a "Will to Freedom" from itself, even if it seems, ironically, to request nothing more than play through uncritical adoption.

As of yet, *Interactive Brainlight* has been staged at festivals and museums in casual and playful ways. The goal seems to be to evoke nonthreatening emotions that contribute to celebrations of neuroscience–art collaborations while also, pragmatically, exercising the limited capacities of EEG to creatively reveal when participants are feeling calm, excited, or focused. Accordingly, the full potential of the work may not yet have been realized. Re-staged as a mental illumination of people viewing more complex social problems, such as incidents of racial injustice or controversial political protests, the work—or one similar—might, more forcefully, foreground what I read as an underlying Will to Power and Will to Freedom.[101]

Intensifying *Interactive Brainlight*'s Will to Power would mean retooling the work to encourage participants to forcefully instantiate new mental projections and to then resist domineering ideological forces, i.e. "overcome" the grand narratives or cosmological or patriarchal perspectives through ridicule (beta), bodily meditation (alpha), and/or logical critique (theta). Here, the three brain waves open themselves to critique and complexification. If allowed to illuminate mentalizations tied to the deadly serious wherein exposures of affect and the emotions put people in danger or produce condemnation or rally forms of solidarity, then the work might better reflect Nietzsche's concept that Being is always a "being-on-the-way," an expansion of power but also a revision of life conditions. The Will to Power, as a self-reflective shaping of one's self to resist the domineering Will of Others, once intensified and centralized in *Interactive Brainlight*, would shout to a collective audience to exercise Will as resistance.

Intensifying a Will to Freedom would require seeing *Interactive Brainlight* as implying an internal "'game' of power relations" in the battle over human affects. The "game," so we might learn from *Interactive Brainlight* as critical-cultural experiment, is about using disruptive or troubling stimuli to reconfigure one's bodily productions—those sudden affective states that are difficult to control. Reimagined thusly, the work might prove more fitting to a cultural landscape rife with identity politics and political noise, especially in so far as the work would then position affect states as a means for rethinking the discourses embedded in us. In the museum context, the idea that *Interactive Brainlight* is in any way an image of "a brain" or that it illuminates "the power of your mind"[102] might be a good subject for the first interrogation. Solely deployed as a means to make audiences swoon while musicians wear EEG sets and play classical music misses something important regarding what the work can say in relation to a controlling neuro-society.

Even if not explicitly critical of social configurations or of popular neuroscience, *Interactive Brainlight* still holds the potential to initiate valuable appraisals of neuroscience. Because *Interactive Brainlight* shares the results of a participant aiming to achieve specific brain states with an audience who witnesses the results, the work allows for a collective interrogation of neuro-centric discourses so often failing to explain what neuro-technologies can actually do and failing to distinguish meaningfully between the brain and Self and Others. Any disconnect between the sculpture on the one side of the room and the facial expressions of the participant on the other side would reverse the feeling that neuroscience's tools are powerful interrogators delineating every detail of the mind through brain activity. The work might, rather, expose the vagaries of neuroscience's tools or the simplicities of their functionality. Three, rather ill-defined, brain states displayed in child-like primary colors do little to convince people of EEG's capacity to instill complex articulations. Neuroscience risks coming off as an elementary school play activity where the kids try out high-tech devices for fun between Basic Math and nap time.

The most immediate benefit of the work may be how the establishment of shared experience functions to strengthen social bonds. Indeed, one undeniable effect is a shared self-exposure wherein audiences gather around and decide, quite delightfully, whether or not they do effectively read the facial expressions and body language of the person wearing the headset. Here, audiences have an opportunity to laugh, to be surprised, to cheer for the participant, or to share in the same feeling. They can believe that they, too, would be lighting up blue, red, or green if wearing the headset. In the act, people feel closer together.

Within this ritual of social bonding, the neuroscience takes a backseat. In fact, we might be tempted to believe that the importance of the work is in the thrill of it, not in how affect categories described through EEG technology pave the way for revising/controlling states of mind. Who really cares about that? The more pertinent "game" of *Interactive Brainlight* is its group dynamic. What seems to matter most to giggling observers is how they feel when they play with the thing and entertain their friends. What stands out is the moment that they leave the room, holding arms a little more closely, sharing a moment. If this is an accurate portrayal, then *Interactive Brainlight* is a technology of positive social relations, not too far distanced from a social media site like Facebook. And thus, it may also be not too far distanced from the shaping forces of Capitalism in a new media age where shareability and likability guide consumer behavior and where the affect of the performance holds significant sway in whether content has any hope of being seen.

But whatever one's view of the work, it seems fair to argue that the sheer notion that neuro-technology acts as effective mediator for us to influence our brains indicates something important about neuroscience: the practice, like *Interactive Brainlight*, prioritizes Self-experimentation and an expansion of self-directedness. But along the way, awareness of the social situation and what it demands must be noticed and taken into account for the artwork to function well. The same could be said of neuroscience. In this case, the way that the work puts an individual in the center of the room "on the spot," so to speak, and situates their vulnerability as playfulness and encourages audiences to cheer highlights the demands of a pre-existing sociality. *Interactive Brainlight* will rightly not allow us to think sociality separately from the expansion of neurotechnologies. We might then begin to see neuroscience tools and experimental practices as socially determined from the get-go just as much as needing to accommodate social needs. In this, *Interactive Brain* achieves an important task: recognizing the co-constitutive nature of social elements in scientific and technical things. It does this—unlike more formal critique—in the friendliest possible way.

Scan 12: brain art is overexposure

If looking and being looked at arouse a range of strange pleasures, as Freud proposes,[103] then brain art, conceptualized as a self-gazing and self-exposure, can be considered an object composed in some kind of relation to (or as a response to) a subterranean psychological compulsion to see people naked and to stare at their genitals, i.e. what Freud calls scopophilia. In this admittedly uncomfortable connection, bodily scanning—and brain scanning as a means to uncover a secret interior—necessarily entails the fascination of seeing more and deeper. The idea is that bodily scanning cannot operate, in the Freudian schema at least, separately from a repressed sexual desire to see the taboo. This desire is formed, Freud suggests, when the child learns of excrement and seeks to uncover things that the parents want to hide (naked bodies, sex, etc.); repressed feelings then build a desire to open closed exteriors.

Although the Freudian proposal might seem overgeneralized or far-fetched, the cinema and the film camera have been repeatedly theorizing in this way, i.e. as motivated by a subconscious drive toward superior presencing, a technical effort to pierce or unveil what goes unseen. The cinema and Hollywood's relentless march toward ever more shocking material is framed as an enactment of some bodily drive manifesting in the technology itself. Following the discussion down the road a little bit might be beneficial to thinking about the relationship of brain scanning to brain art and the fascination of seeing brains in places like museums.

Film theorist Monica Michlin details the relation between the film camera and the Freudian proposal in her article on "exposure." As she explains, "Exposure in its double meaning lies at the heart of the cinematographic experience ... the tabloid photographer's camera flash, a star stepping into the spotlight, the brilliant whiteness of a hospital corridor, sunshine at noon, reflections off snow," these are "overexposures" that "open the eyes wider" and operate "metonymically and symbolically within the film," serving as amplifications and representations of a desire to see too much.[104] The film camera itself, in terms of its technical functionality, also can be read through this Freudian vision; the cutting together of a film produces a stream of looking into another's life with binoculars. Likewise, the focus makes quickly clear what has for so long lingered without recognition in the fuzzy (subconscious) background. The ZOOM performs the invasion of intimate space from a safe, often hidden, distance. The opening scene in Alfred Hitchcock's *Rear Window* illustrates in live action the specular fascination of film, operating, as another theorist says, as a narrative making "conscious use of Freud's subconscious."[105] The storyline can even compel us to turn around and look behind the camera where we see Hitchcock himself peering through the camera's window.

The brain scan, we can now safely argue, ventures to allow sight much further into the body than the film camera. But whether the brain scan produces the same kinds of pleasures or compels the spectator to perform the same fascinated looking is another question. Brain art (or broadly, those works showcasing bodily scans) help form an answer in the way that they relocate the brain scan. Once outside of the sanitized scientific laboratory and in the museum—i.e. into the realm of public viewing and fascination— the brain scan is no longer safe. The psychological perturbations of the penetration cannot in brain art be sealed off by the authoritative closed doors of the laboratory or by technical descriptors. Context makes all the difference to psychological botherings.

Like the movie theater, the museum space exhibiting brain images can be understood as the inevitable place where such images must go. The very design of the place enhances the drive to expose. As Mulvey notes when speaking about the movie theater:

> The extreme contrast between darkness in the auditorium (which also isolates spectators from another) and the brilliance of shifting patterns of light and shade on the screen helps to promote the illusion of voyeuristic separation. Although the film is really being shown, there to be seen, conditions of screening and narrative conventions give the spectator the illusion of looking in on a private world.[106]

Something similar might be said of the museum. As with the movie theatre showing a film about private lives and sexual adventures, the brain of the visual art world can be understood as inherently voyeuristic. The display of strikingly sensual depictions of the brain in an artwork is a bodily invasion hospitable to interpretation and made inviting—some interesting commentary on the artist's or the visitor's experiences perhaps. The figurative and literal insides are hung up there on the wall to be seen while the environment of high culture lends force to the illusion that social taboos are probably being addressed here but likely no line is being crossed in the display itself. Privacy for the person pictured is being maintained, we feel assured, by the institutional rigor of the space. Likewise, the implied universalization of the brain protects against unethical revelations of psychological specificities that George, Mary, or Coleen might have entertained on a Friday night. This safe space mentality is carried along by the context even as brain art gets much of its allure from proposing that the most secretive of all of the human interiors—those creepy or strange or wonderful mentalizations—are being flat-out exposed. Colors showing brain areas firing tickle the viewer and encourage suggestive decoding. One might well mark the person's brain as revealing the sicko or the hidden psycho. Why else would we look? What makes us scream with delight at the brain image?

This gets at distinct differences between the movie theater and the museum. Unlike the screening of a film rife with lurid exposures, the image of the brain constructs a voyeuristic sales pitch that cannot really be underwritten by the product. The images do not deliver on their promise to penetrate the Self even if they penetrate the body. They remain a little obtuse. The viewer is not sure what is being presented. Pretty colors. Some shapes. A scan. So what? … *Oh, yes, yes, I see the title of the artwork now. That must be a psychopath! Wow! Look at that!*

The bubbling up of any satisfaction, including any eroticism, from seeing a brain scan image relies to a greater degree upon an overlay of imagination. In lacking the immediate punch of nakedness, as it is usually encountered, the brain scan does not elicit the same effect as a film with titillating detail. What the brain scan produces is a *making-naked*, at best, or a momentary feeling of meddling in another's secrets. But the image is a metonymy, a standing-in for a body and for a Self, nothing so literal as an actual naked body. Yet, funnily, the scan so clearly asks us to see the flesh of the Other; the brain scan composes another kind of nakedness entirely.

Because museum spaces have the characteristic of displaying numerous images near each other, curators have the option of putting two works in such a proximity that they cannot be seen in isolation. Since a brain scan is such a strange and compelling art object, we might take the opportunity to try this out to see if we can extend the interpretive possibilities. And to provide some humor and a little fun, we will work with Freud. That is, we can imagine putting the brain scan image right next to, say, Lucian Freud's "Benefits Supervisor Sleeping." This nudie juxtaposition intensifies, perhaps, the bodily fascinations of each, and it is a convenient choice for recalling Sigmund Freud's theories (Lucian's grandfather). If we see both works together, then we find a way to ruminate on their different approaches to bodily representation and different normalizations of exposure.

Let's give it a whirl.

On one wall, we stare at Freud's painting. He reframes nakedness as high art because he paints rolls of fat in a mundane way and thereby sets nakedness, and not the more figurative "nudity," as means to understand a contemporary life. He, thus, participates in a long and varied art history tradition of turning to the common for social insight and criticism. Looking at the adjoining wall, we see the brain scan; we will imagine that it is titled "Self Portrait of a Benefits Supervisor." As an artwork, the image functions in a similar fashion. Both pretend to grant the viewer some privileged access into the subjectivity of the person depicted in the image but only through the encounter with stripped-down flesh as representation. Both certainly ask us to rethink what is taboo and how far the creator of images in our culture will go to put viewers in touch with those bored, overfed, under-stimulated people abiding in our suburban landscapes, which make a difference and require recognition because they tell us something about our society. Yet, the feelings being generated in the two artworks are nevertheless manifestly different.

Freud's painting, "Benefits Supervisor Sleeping," elicits surprise or bewilderment;[107] the brain scan, on the contrary, appears aesthetically charmed, so much so that viewers risk falling in love (perhaps with themselves) with the bright colors and symmetrical patterns of neurobiology. Lucian Freud does not seem to want us to fall in love with the Benefits Supervisor. Take a look at the painting. She does not see us looking at her, for one; she offers no enticement as she sleeps. She does not welcome us forward. Thus, we become the voyeur-as-painter dipping our brush and splotching her every fold of fat onto the canvas as she lies there unconscious. The image, then, could be framed as inherently a patriarchal one, since in viewing it, the audience is thrust into Lucian's own position, forced to stand in an unsteady ethical relation with the Benefits Supervisor who sleeps. In this way, the painting asks the viewer to assess the power relationship. Who supervises whom? Who is in control? Is the painting completed with permission? Is the looking that we must do here a "male gaze" that objectifies the Benefits Supervisor? Or does the painting merely ask viewers to consider the woman's body for what it is—raw, unglorified flesh? Perhaps we are being asked to celebrate what is denigrated and to elevate what is derided. This is an assault, we might convince ourselves, on traditional concepts of beauty. Even so, we wonder if this is the celebrated artist-man's job to elevate or deride the woman's body. We blow out air. The painting, we then think, is an erotic compulsion, a naked fantasy, excuse the pun, rooted in a neurosis of looking. Or maybe we get stuck in an interpretive immovability, just there, staring and not looking away. We are, in this case, once again slipping into a neurosis.

In contrast, the image of a brain network laid bare avoids such tricks because it is, we are prone to think anyway, a picture of us. Even if the image is not us, exactly, then this image is enacted with permission, unlike Freud's. But that is when we start thinking about it more. The brain scan is taken from whose perspective exactly? Are we seeing what a scientist sees? And could we know if it is shown here with the person's permission (whose brain this is, after all)? Did the scientist or artist ask the person before transforming it and transporting it to the curator? What kind of looking, then, are we doing?

If the brain scan enacts or initiates a male gaze, it is likely to be because neuroscience is so dominated by men. Or we might rather suspect that it is an image of patriarchy because the act of making bodily representations in art remains so overindulgent of male subjectivities and significations. Suddenly, we remember that the image asks us to view our Self-as-brain when seeing this other brain: then we wonder if this is actually the brain of the artist, or maybe it is offered as a social compilation, a composed brain. Either way, we sense that we are encouraged here to look for something that makes us feel unsure or queasy: an image asking us to overwrite ourselves. It upholds a penetrating gaze not far off from what Donna Haraway once described as a gaze that "mythically inscribes all the marked bodies, that makes the unmarked category claim the power to see and not be seen, to represent while escaping representation."[108] Yes, this neuroscientific image with an unseen maker (a whole set of actors, human and nonhuman, actually) who escapes representation constructs a male gaze after all, we can now decide. It is a gaze that invests in supremacy and extracts itself from "situated knowledges," implicating the gazer as the one that must repress any reliance on a body or an outside, just as the scientist must repress any social or technological reliance in the act of its creation in order to transform it into an objectifying scan and then into an artwork that can just willy-nilly show all of us.[109]

In being invasive, and in this special way, namely, in purporting to show a material core of others and our Self simultaneously, and doing so in public, we can gape at it with surprise, horror, and/or pleasure. But to see the Other on display as a brain, we must also transpose the looking onto our own bodies. In watching those pleasure centers light up in red and blue across common reward circuits in the amygdala and nucleus accumbens and then feeling some pleasure in doing so, we sense that the artist implies that what we are doing is something like enjoying ourselves in public. If the drive to bodily exposure, or overexposure, is there at the heart of the brain scan's effect when producing excitement, then we might have trouble discerning how that feeling is much different than the deviant one. Perhaps it is a matter of degrees.

If enjoying the pleasure of a brain scan is different than the (Sigmund) Freudian pleasure induced by witnessing the taboo-breaking boldness of blunt nakedness, then it may be because the penetration goes to the heart of the Self, and in that sense, goes deeper. The representation sold as an image of the secretive inner life crosses right away into a forbidden—Self-exposure, yes, but also Self-destruction. The brain scan is a deconstruction, a thread running contrary, a signifier (the scientific scan) declared stable and non-narrative in nature even while the signified (Self) is declared false and narrative to the core. There, the intelligible scan overwrites the unintelligible of the Self, resituating phenomenological experience as a self-trickery being completed by a self-evident neurological mechanism.

But to try to eradicate the Self in this technical way is an impudent incursion of one stability onto another under the guise of a revolution against stable forms. Accordingly, the effect is ironic. The protagonists pushing up against an objective or universal Self do not recognize that they do so by functioning in the same way as the antagonist, even if they do it with different means. In other words, the scientist with the scan might protest against the metaphysical for the same reason that the metaphysical is being disregarded, i.e. for offering an enculturated and unshakable Truth about existence or Self-Being that does not recognize its own enculturation. Thus, the new path for re-understanding the Self (the destruction of ontological stability) is only the guise of a revolution. The real revolution would be admitting that neuroscientific exploration follows an enculturated narrative and set of interpretive possibilities, ones which may be viably integrated, if desired, into worldviews invested in stable forms and metaphysical conceptions, absent any assertion of material totality and cultural dominance. Revolutions can be more tempered than we might first imagine.

If the attempt to eliminate a Self-concept is seen as only a Self seeking to perpetrate a cut or a violence against its Self with a brain scan, then it is a kind of acute aggressive-eroticism mixed with an egocentricity. That is to say, it is, ironically, a Self-obsession failing to recognize the act as such. Here is where the strange title "Self-Portrait of a Benefits Supervisor" names not only the brainy artwork but also the spectator.

Once understood as a severe psychological incursion, a brain scan made into art is a plea to heaven or a ride into hell. Perhaps it is better to phrase it this way: socially, we can understand brain art purporting an exposé of the Self as fitting within the Judeo-Christian tradition—and this is not surprising, since meditations on the Self are often infused with religious signification. To play it out: we can say that the revered art space can be imagined as a church; the brain scan image could be the picture of the naked human standing without the little grape leaf covering, proclaiming that its secrets have no shame anymore because there are no exterior dependencies. As when reading the Biblical text, the audience must then look at the nakedness and repent for indulging in such remarkable insurrections against an Order of the divine, or the audience must cherish this deviance as a revelation of divine secrets or, rather, of secrets stripped away from anything relegated to the divine.

In the case of the former, visitors understand brain art as a critique; they note and listen to the corrective psychological function that regulates deviant behavior. The neuroscientific enterprise becomes the devil's realm, a kind of temptation that looks way too far inside and sees way too much. Brain art partakes of the fruit of the tree and gives knowledge of our nakedness.[110] In looking, in eating, the visitor then repents and turns away. The brain scan cannot be fully adopted nor worshipped, for it leads resolutely away from a better path, away from an ever greater complexity of internal sensations that forge the experience of the body, emanating from the ecology and, per-haps, from mysterious forces far beyond human comprehension or material awareness. To not turn away is to invest in the brain as light and to say its exposure is total. Such a position relies on the notion that there is no external, no gods. The position rejects various manifestations of what goes unseen as integral and residing inside of the human. To do otherwise would be to stand resolutely against something real that humans appear to feel about them-selves in their weak and lonely state on earth, some ungraspable, some undefinable that yet gives definition.

In the case of the latter, however, visitors understand brain art as con-firmation of neuroscience's power; enshrining the image in the museum is understood as a celebratory and liberatory exercise. The congregation par-takes of the fruit of the tree precisely because the devil does not deceive but, in some essential way, tells the truth—more about our condition on earth can be known when we taste the forbidden fruit and "open our eyes."[111] In looking, in seeing deeper into the body, visitors feasts on themselves. The Self is right there, nothing really, certainly not mystical, and to be denied as essentially and as strongly as any religious essentialist might well do in the reverse. The fruit of the tree of knowledge tastes like flesh and blood, but if it is a last supper, then Jesus is not in attendance.

In thinking through this weird combination of Freud (meaning both Sig-mund and Lucian) and the Biblical narrative—that is, in entertaining the idea that the seduction of brain art in the museum emerges from some deeper subconscious indwelling that can initiate reactions described through concepts like exposure and the heaven–hell dyad—the appearance of the brain, as a dissected mind, can be likened to unveiling the secrets of human-ity; but these secrets, as the devil says to Eve, were meant to be kept hidden. The brain, accordingly, transforms into the divine image dwelling within the very flesh of Adam. The neuro-artist plays the role of the priest/ ess there to perform secular rituals in order to mediate our relationship to what hides but what dwells deep within, a reality that gives us life and is felt, but remains in the realm, most basically, of the mystical until it can emerge in another form, be presented in the light, be made into a tangible representation, which explains what we are.

Once the power to reveal an objective truth of human life is handed to neuroscience, and specifically outside of any social histories intruding over the interpretations being given, the God of mystery dies. But God rises again precisely, and ironically, in acting as the neuroscientific One designat-ing and delineating bodies. What a thrill when neuroscience's revelations guide our narratives; indeed, the scans can be put into a museum precisely because they are like the directives of a god, easily situated as showing us ourselves. Through them, we dream of divine realms. This is what museums are after.

No wonder Caravaggio painted "Mary Magdalene in Ecstasy." Upon seeing God, feeling His closeness, Mary yearns, sucking in breath with near-orgasmic pleasure. The dirty sexual body is transformed in the image and presented as part of the divine order, indeed united with the divine; old divisions between the lowly sinful flesh and the elevated or redeemed spirit are rocketed back together. They appear as two lovers reunited after a dispute. The hint of erotic pleasure makes perfect sense—the bodily is made in the image of God, and there is something heavenly about it, Caravaggio seems to say. Clearly, he does not miss what the feeling of God in our flesh can do.

Of course, some art critics sneer at Caravaggio's depiction of divine pleasure and say that the model that he used was nothing but a common whore; thus, what he paints is carnal and typically depraved.[112] But to adopt this line of argument is to indicate that behind the scenes of any glorified representation is a set of social histories and enhancements that take should precedence when the representation offends. The wonderfully idealized then becomes an image of the artist's own gross bodily needs or of the privileged class taking advantage of the destitute. Caravaggio is but a whore himself. But that is too easy. We cannot miss that Mary yearns in the image, regardless, and we stand today in the museum, regardless, wondering about her posture and facial expression. The image has an emotional charge that can push us to think more about the Biblical text, the role of theology in our lives, or the meanings that we attribute to our various desires. In like manner, although it may be intelligent not to invest too much stock into all of Freud's theorizations, we cannot totally overlook how deeply the brain image can focus our attention on our yearnings and various emotional states of existence. Like Caravaggio's work, the brain scan illuminates excitements and bodily pleasures. If we indulge a little in the scientific self-inscription, then it may be because we cannot disregard the feeling that the image incites us and excites us.

If brain art is "overexposure," then it is because we want to open our eyes and see more. Who we trust to do that and how we go about it makes all the difference. We need a Caravaggio of brain art. But more so, we need a Caravaggio of neuroscience.

Returning to order

At the intersection of the body's mystery and its aesthetic illumination in art, another realism comes into focus. Like past realisms, Affective Realism pursues truth and verification, solidity and confirmation, the psychological "real," the grounding of the body's everyday experience, scrupulous attention to detail—but this realism obsesses about the hidden inside. The farmhouse, the peasant, the lovely bowl of fruit say much but much less, in fact, about the social real than the frontal lobe says, or so we are compelled to believe. The claim is photographic in nature—the neuroscience image cannot lie. Of course, images lie in all kinds of ways, we know this. But the rhetorical move is reminiscent of a "return to order" from the postwar period of Italian Realism, which emphasized stability after psychological turbulence,[113] leading to the neorealist obsessions with technique and hybridization.[114] Apparent in some brain artworks, this Modernist project resonates within the impulse to fold biology over top of sociality.

If there is a new escape from disorder, then it may well be trying to escape from the post-structuralist disorder of semiotic scaffolds ("turtles all the way down"); or it may be evading the disorder of technological complexity; or it may be trying to reorder the raucous disorderliness of shaky institutions like NATO and the EU; or it may be an escape from exhaustion in regimenting our workplaces and lifestyles. Whatever the case, and there are many cases to be made, a new realism—Affective Realism—reimagines the human condition, and so broadly, through portraits of discrete brains. It operates, on that score, like the Italian Realist re-imaginations of national structures within the tiny scenes of everyday life painted with attention to every spot of dirt and mold.

The goal of Affective Realism is to sniff out the real of the common masses, to discover what lies beneath the surface, and to do so with intense attention to the micro of material structures. But heightened attention to technical innovation amid images of everyday people to outline their suffering and grope for some salvation, as the Modern Italian Realism project attests, risks several different backlashes: first and foremost is the refusal of the notion that any individual is actually represented; second is the recognition of alienation from the privileged structures that seek to capture those being alienated; third is retraction from the idea that an order can be imposed onto such dense complexity—much less through systems which exist beyond the voice of those being ordered; fourth is a disgust in the fact that anyone seeks to profit or make pleasure from giving attention to the emotional components and strange features of human beings.

Each of the above backlashes can be watched in the Italian films of the 1960s appearing after the postwar period. By renewing a focus on "mystification and illusion that keeps its occupants in a state of untruth," these films push back against the 1940s. They are visual exercises in unmasking epistemological, hierarchical statuses as lies, wherein they aim to perform a "turning away" from the notion that the world is simple enough to know anything. These films worry about the real-ideal concept implied by Plato's cave, specifically in relation to the epistemological power of the production of visual imagery, even as they reject the idea of an ideal to be located and grasped.[115] As Millicent says, Italian films of the period after Italian Realism (1960s post-realism) make quite explicit "the anxiety that has always accompanied the mimetic arts, from Plato on, with their power to seduce us away from the path of true knowledge." And then Millicent offers a telling reflection on film's technicity: "Cinema is ideally equipped to bring that anxiety to a head thanks to its technologically driven claim to reproduce 'the real.'"[116] The same can be said of the fMRI scanner. The question, then, is whether we feel more aligned to 1946 or to 1966. Brain art might sometimes prefer the 1940s in as much as it can offer a kind of macrovision about human behavior through microvision without intervening layers, but it can also get groovy with the 1960s. In the 1960s, the cinematic effort to "deconstruct the world of beautifully, artfully posed, extravagantly costumed, heavily made up, eternally young women," encourages viewers to reject the incentives offered by perfect worlds. The fashionable characters in the 1966 film *Blow Up*, for example, cannot find solace, and the closer the lead looks at his photographs, the more he sees how complicated and yet false his world has become; ultimately, the film foregoes the "happy ending."[117] The final scene shows the creepy play of mimes in a city park, and the truth of a murder in that park remains elusive.

A creative depiction of the brain using neuroscience's illuminating techniques intends to penetrate a foggy mind and shine a spotlight down onto the core driving human desire, imagination, creativity, and personhood. But anyone who has lived in Appalachia knows that turning on the bright headlights in the deep of the fog is not always a good idea; it blinds the driver. Affective Realism, when expressed as a brilliance, shines too brightly for its own stage. Affective Realism does not open the eyes wider in its overexposure, it more often gives presence to the blank white laboratory room walls with their overwhelming sanitation. In some cases, the artists mentioned in this chapter, of course, intuit how the too sharp a focus on instruments and authorizations can complicate the steering, so they tap a little on the breaks. But in other cases, they seem to speed downhill, flying around windy curves with the lights blaring extra-bright, hoping to reach the top of the mountain. Buckle your seatbelts.

The transcendences of The Brain

Affective Realism, as a thematic in brain art, gains historical and philosophical perspective when entertained in reference to what Nietzsche claims Christianity achieved socially by offering a universalizable real world, i.e. what Nietzsche calls "the Christian hypothesis." Amid the contemporary loss of many of the advantages to the Christian hypothesis, the brain holds the potential to take up the proverbial staff. Seeing how this works, or trying to make the argument, at the least, helps us to consider how Affective Realism might appear to be desirable and develop out of a train of thought that Nietzsche inaugurates.

According to Nietzsche, Christianity functioned well because it effectively

> bestowed an intrinsic value upon men … it granted the world a certain perfection despite its sorrow and evil. It also granted the world that proverbial "freedom" [from sorrow and evil] … It assumed that man could have a knowledge of absolute values, and thus granted him adequate perception for the most important things … it prevented man from despising himself as man.[118]

Today, after the reign of the Christian kingdoms is over, as they were once known anyhow, it must be observed that those emerging from this tradition adopt only an uneven selection of the reasons that Nietzsche lays out: (1) Intrinsic value for living beings; (2) Objective perfections; (3) A way for freedom from sorrow; (4) Knowledge of absolute values; (5) Belief in adequate perception; (6) An unassailable reason not to despise the human so sullied with small-mindedness, division, hate, hierarchy, and selfishness. Despite many scholars asserting that Capitalism works to fill these hypothetical absences and in so doing expand and perpetuate its own system for the benefit of the elite,[119] neuroscience may be better suited to the task.

In being material, the brain reverses the means of human transcendence, coming back down to earth, so to speak. It serves much the same function as the Christian hypothesis but returns the longing for solidity and self-assurance back to itself. The result is that each subsequent point of the Christian hypothesis finds new grounds stronger than that offered by most other things, namely Capitalism. A basis for value endowed to the living, a means toward perfection, a way to freedom, a clear set of values, and an adequate perception are better secured with attention to the neurobiological.

To draw out the argument, the first observation is that the brain (once tied inextricably to mind) makes for an immediate way of evaluating and re-inscribing what it can mean to live with stability and security. Capitalism, in contrast, dislocates the Christian hypothesis; an ability to achieve perfection and a sense of escape from sorrow, for example, is tied to consumer behavior. The idea is that consumption of alluring objects appeases existential unease. A big screen TV or new computer establishes "the (valued) person" within a broader system of Capitalist values, and people can feel its effectiveness when the neighbor celebrates upgrades with a pat on the back and a hearty smile. Invitations to high-rise cocktail parties assure consumers that they are on the track to a meaningful life. Capitalism functions to stabilize social positionality, and acquisition gives assurance of survival amongst antagonistic Others, simultaneously displacing and intensifying the deeper (Freudian) desire to gaze at others and be ourselves viewed. Is it any wonder that touting Prada down Madison Avenue satisfies so intensely or that it palpably enhances our allure and embeds in our fantasies of becoming alluring to Others? We need not marvel at the way that Internet apps of self-exposure or drones whose cameras spy on unsuspecting dotes at the beach fly off the shelves and sell themselves. In these consumer objects of voyeurism—which all consumer objects are at their core—we discover short-term satisfaction in pretending control over others and in ensuring Self-stability. Yet, at the same time, we can do so without destabilizing a social hierarchy, one which enables *the feeling* that we are "winning," making social progress, achieving degrees of perfection against an imaginary Capitalist measuring stick. To repeat the feeling, we buy more.

But the satisfaction almost immediately dissipates. We hear Mick Jagger wailing, "I can't get no satisfaction," over and over again, as if the repetition will force a realization of the desire.[120] The attempt to build meaningfulness without the possibility for real change in ontological conditions is frustrated by spending all of our savings. Yet, the brain, expressed as a *kind of* brain, holds the conceptual advantage, even though this brain, too, requires consumption to be properly exercised as a new transcendence. Our "first nature," so to speak, can be perceived as offering salvation from personal unhappiness above and beyond what Latour calls our "second nature," namely, Capitalism.[121] The core of the problem, if biological, cannot be resolved by adding layers of plastic, silicon, or gold.

In his discussion of Capitalism being inherently affective and emotional, drumming up "miseries and luxuries" in the endless drive for greater riches and the predictable failure inherent to continual achievement, Latour jokingly says, "apparently it is the laws of Capitalism that Jesus had in mind when he warned his disciples: 'Heaven and earth will pass away, but my words will never pass away.'"[122] The joke is fitting for a world with mega churches raking in big bucks but also for a world where billions of dollars go to corporations promoting reliance on fossil fuels amid debilitating pollution in the oceans and atmosphere. The earth may well pass away to continue the life of Capitalism. However, if the banks survive the end of the world, it is not because of their strategies of appropriation and enthusiasm for ensuring the proletariat remain pacified while giving all of its time and labor to an elite class that benefits. Rather, the words of Jesus (his promise of salvation and our transcendence) continue because both the elite class and the proletariat live with/in bodies that cannot but recognize an existential condition: a longing for "a transcendent world beyond," as Latour says.[123] The world may die a rotten death because of Capitalism's destructive effects, but Capitalism is not somehow a different case; it is the same case.

Although Latour sees the economic replacing the salvation of the religious or metaphysical in making his joke, I see instead a physicality operating within the affective response to Capitalism: a neurobiological accretion. With such a bio-centric characterization, I turn my argument inside out. I critique the existential resolution—salvation now offered through a transcendent and abstracted The Brain made accessible through social manifestations of neuroscience—precisely by pointing to the body as the instigator of the problem. I humble the materialist replacement of the Christian hypothesis but then look for neurological ground from which to understand the psychological dynamic. In making this gesture, I become to some extent neuro-fascinated myself. I perform a self-gaze where I am the object and the subject. This is the power of The Brain, and the heart of the point is made. The neurobiological stimulation that participates in driving me to want more out of life—to expand my own life—can never entirely pass away, even if The Brain, as it is known, is able to fulfill some promises for us.

If we choose not to turn to Capitalism or to God, then where else can we go to satisfy the need for more and to build some resolution to psychological longings? Perhaps we will choose to love The Brain—its reference point (the brain) is inescapable, after all. The Brain constructs means to grapple with our dependencies and insecurities. But we are not likely to find a quick and easy paradise.

In adopting The Brain to ensure self-governance, we must note that the solution contains within itself a fundamental interior loop; the organ cannot think its way out of its own condition nor resolve itself any more than a knot can untie itself. If the brain remains in control of how it is being interpreted, paving in its own limitations, then we seem to be running in circles. Around ourselves we go, merrily, merrily. In that moment, we risk falling prey to obsessive self-fascination. So we might try to learn from the legend of Narcissus' adversity, i.e. how he starved himself by spending night and day staring adoringly into his own reflection. Of course, there is no indication that Narcissus felt his inability to escape his own self-gazing a serious problem while he was doing it; we hear of no hunger pains or tears as he wasted away. By all accounts, he was chuffed with himself right up to the end. That is so very human, isn't it? But, certainly, he was delusional, trapped by the beauty of the mirrored image that was himself.

Might brain images encourage the same happy repression? There is, after all, a moment of contentment when purporting to see ourselves in that technical yet colorful representation. But those same representations can be re-seen as a reminder of an ontological trickiness. They show us an involvedness of matter in the process of making a thing known.

But embracing the difficulty of dealing with matter as we set out to resolve questions about this world—most pertinently how to keep the body running—is not to encourage giving in. We need not deny that more must be done to improve human life and, yes, to learn about the brain. In fact, becoming aware of limitations and of absences motivates, even requires, clinging to a hope for all kinds of futures, biomedical ones certainly included; we "fix our eyes not on what is seen but what is unseen," as the Apostle Paul says.[124] That quote centralizes the lack of present options for resolving the biggest questions; it reflects a metaphysics, sure, but also a not-knowing; in it, we detect a lingering background sense of alienation or dissatisfaction with living, namely total investment in a hope not yet seen. One works to see that hope, even in the Christian worldview.[125] And the central positioning of that hope is exactly what the Christian hypothesis likely intended from the beginning—but the unseen and undetectable part embedded in hope is what Nietzsche does not consider when arguing that the Christian Hypothesis's universalizable real is a historical contingency needing to be replaced by humanism's discovery.

Reading through the narratives in Exodus helps to elucidate the idea. From the first few chapters, one walks away with a view of people living with recurring sorrow while yet depending on a faith that was uncomfortable and alienating. The faith of the Israelites entails a willingness to not fit-in to the social norms of Egyptians and to flout human orders due to the awareness that an Almighty exists and disapproves of the one who could grab them by the arm and then cut them to pieces, namely, Pharaoh. But here is the more pertinent and interesting part of that faith: a hope that God will care. If Nietzsche is correct in saying that Christianity provided for basic psychological assurances and more internal security, then he does not try to explain the actual behavior of those held up as models of the faith. The point is worth teasing out in relation to The Brain because it demonstrates how Nietzsche's way of thinking about what *should be* misses how the inescapable—the unseen, unfelt, and unknown—remains a force and is foregrounded in many faith-based worldviews. The connection to The Brain will, hopefully, be obvious.

In brief, the faith concept presented in Exodus retains an elusive, mystical element, adopting a bold but basically unproven acceptance of a benevolent immensity in the universe (they were enslaved for 400 years after all). Nevertheless, the Israelites show a waking fear of relying on one's own knowledge, a kind of trembling about incursions on that immensity. This is evident, for example, in the refusal of the midwives to listen to the all-powerful Pharaoh—who could certainly punish them and their families—due to their trepidation of transgressing against God, who at that time, interestingly, had not yet fully revealed Himself to the Israelites as "I Am."[126] The name of God given there itself is mysterious. So the faith narrative, from this angle, suggests that humans might do well to side with the unknown mystery of things over the known circumstance. The known in this case was Pharaoh's desire and reputation; the unknown was whether God was really there and would care or intervene as well as whether Pharaoh would kill them. Living in that type of darkness and often against one's own interests (as in Exodus) is no easy matter. And that mystical aspect to how things will end up, that trust in an ethic or an intuition, is quite clear in Exodus but not evidently what made Christianity successful over the ages; thus, Nietzsche can, perhaps, more easily overlook this aspect. The successful spread of Christianity in history relied on displacing the humility and the frightful unknown built into passages like those about the midwives in Exodus. I don't think it is very controversial to suggest that Christianity appealed because it provided assurances but also because kingdoms of past centuries located compatibilities between the expansions of Capitalism and colonialization with the church's regulations and formal doctrines, and those are the aspects that Nietzsche clearly detested.

It remains fair to say that Nietzsche correctly noted that the Christian hypothesis resolves some of the body's vulnerabilities. But advocating full acceptance of the body while yet wanting to avoid securing stability from an outside source would mean—if I am reading him correctly—mentally overcoming any oppressive, circumscribed social condition. The position contains an over-reliance on knowledge of one's self, just as it does on knowledge of one's condition and future condition. Equally, it appears to depend upon a strident, exceptionalism and individualism as much as on the mind–body dualism. So if Nietzsche blames historical Christianity for foreclosing human potentials, then he might also be doing the same by arguing that the "overman's" future comes from an ability to discover cultural constraints and values as time-bound creations.[127] In other words, the view does not allow for an environment where life goes "unseen" or where the "unknown" flourishes (even amid any known). It lacks some of the flexibility, indeed fluidity, of human perspectives and conditions, installing the same kind of superiority it seeks to abolish. This is Heidegger's assessment, namely, that

Nietzsche proposes a "metaphysics of presence" proposing the state of human Being to be static, even as he tries to reject metaphysics. In this way, Nietzsche can commiserate and spin solutions.[128] Here, I want to draw a line of connection: Nietzsche's proposals resonate, in terms of their direction and tenor, with the popular promises of The Brain of today.

Moving from oppression to freedom and unknowing to knowing so often in human history boils down to more clarity of sight and more definition, excluding the "maybes" and the "in-betweens" and the "some kinds" and the "sometimes" and the "surprise" and the "mysterious." The body, as a set of Becomings, or as a collective of the non-human, non-cognitive, and affective, seems opposed to the full and complete; the body mocks dichotomies and unqualified stances. The Brain, however, like what Heidegger says of Nietzsche, wants the human to be One way. It would be easier, wouldn't it, to be superior and self-contained?

The Brain offers an alternative to metaphysics but within something like a "thinking through" or an over-reliance on the ability of neuroscience to discover what ills us. What we get are the absolute best models for learning. Gamified means to expand human cognition. Neurotransmitters for speedy test takers. Here, of course, I am speaking about the caricature of the neuro-craze and its full maturation, i.e. The Brain, relating it to a trajectory of thought that seeks alternatives to metaphysical conceptions.

How successful my above inter-connections have been remains up to the reader, of course. But the grounded point may be simply this: to allow the body to move with the actual science of the brain and to overwhelm The Brain. Looking to the brain, not The Brain, the body feels for another way, and with a performative and active brain it seeks yet another, and another, through this world. We might in elevating ambiguity find a way to re-approach "Nature," to always misunderstand it a little, to see it as a perpetually disjointed expression—a composition of knowing-and-not, of having-and-not, of being present-and-not. There we grant entrance to the impossible, and we trust in what is not yet composed as well as the not yet known as always inside of what is known. The vulnerable feeling, then, becomes not a problem per se to destroy and to overcome but an inevitable part of living.

A popular The Brain wins hearts in its totalizing message of a tangible, visible "There it is." From the apparentness of it all, The Brain becomes like a human god with an earthly advantage over the unseen Christian God of faith. In fact, we can so easily now imagine a future where The Brain reigns supreme: this is a future where we relinquish every last claim of privacy over our mentalizations, where we allow our brains to be delineated and described back to us, where we give ourselves over to the total penetration of neuroscience for the sake of stable governance. To do this, we must be scared to death or be in love. If not, then we would certainly end up despising ourselves.

We would hate ourselves for getting hitched to a neuro-society. We would embody the jilted lover, looking back, saying, "And for what is this for?" Did we marry because The Brain complimented our management decisions or our facial expressions? It would be too hard to pretend that The Brain really does understand us and accurately describes us. Indeed, to shun the belief that the scan sees what is *truly real* about us and to reject its interpretation as the best means to make decisions while still saying out-loud our vows of marriage would risk leaving us feeling used. Here, again, we follow the structure of thought in the Christian tradition: love, proposal, and then marriage. But love is required. And after marriage, we must stay true.

Consecrating neuroscience's soothing promise to ensure our healthfulness and longevity, to provide greater agency and security, abolishes any guilt about the intractability of thinking more about processes of Becoming in neuroscience. The alternative is to run away vicariously with a secret lover called vulnerability. In marriage to the powers of a neuro-society, however, we make abject the Self's awareness of its own ontological missing piece. Subduing that aspect of ourselves to keep up appearances, we perform, as Kristeva insists, "one of those violent, dark revolts of being, directed against a threat that seems to emanate from an exorbitant outside or inside, ejected beyond the scope of the possible, the tolerable, the thinkable."[129] With successful ontological abjection, we avoid that "which is opposed to the I," and by this Kristeva means an opposition that unsettles at our core, that jettisons the rules of our game, that makes uncanny the repression that upholds our Self as such.[130] In falling in love with a neuro-society, bequeathing ourselves to The Brain, we exchange a vow: "to have and to hold, from this day forward, till death do us part." We uphold the "I" because we allow neuroscience to narrate it back to us in languages familiar.

Just as we choose not to abandon our marriage partner, we choose not to critique the neuro-society partner. Any twinkling that infidelity might be possible, even possibly enjoyable, should be crushed. But soon enough, we ask questions about the roles that we have been assigned; we are overtaken by the quick fear that we will not achieve any human transcendence and have given ourselves completely over to another only for our own self-exposure to be tossed aside in an affair. The picture frame where our face belongs, god help us, suddenly depicts another—the head of a security agency or a striking politician. As it turns out, the spouse is more interested in making a career of this marriage then setting us free. And then there is the admission of guilt: "Well, I never really could make good on all my promises; it's me, not you. I'm indirect, limited to correlations, not really very detail-oriented actually, have trouble cleaning up after myself, and I often take other lovers just for the prestige." The Brain as a faithful and tender partner was an illusion. The Brain, as it turns out, is dependent on an outside and unable to provide objective inscription. We therefore choose repression, and we choose the love fantasy. We must! To so totally tie ourselves to The Brain, we must have exhausted economics, exhausted religion, and now feel so acutely exhausted by our impending mortality.

The brain as The Brain appears a more effective return to the Christian hypothesis than Capitalism. It sustains all of the same tenets. Brains can be, for instance, granted inherent value in their creation of sentience; neurochemical balance and pharmaceutical intervention can pave the way to freedom from sorrow; scanning technologies can assure us of equity in determinations of Law; measurements of neuro normality and mass comparisons of many brains offer an objective phenotype; detection of abnormalities function to explain why, exactly, we might despise Others and how we might institute therapies to resolve despicable behaviors. Likewise, we can also come to believe that through the brain we achieve an adequacy of perception. With neuroscience, we can believe that we come to know the Truth better, ourselves better. We can fall in love with ourselves again. Importantly, we can also establish a common ground from which to act. We can discover inherent identifications. We can order human life, now and into the future.

This entire process—the courting, the love-making, the marriage feeling, the long-term dedication, and the hope for a better future—is performed and also interrogated within both of the two themes seen in works of brain art, i.e. Neurosensuality and Affective Realism. These themes contemplate the emotional side of neuroscience and elucidate how we hope to get a hold of ourselves. But if we look closer, then brain art—even those works that exist in strict matrimonial relations with *the real* of the scan—are not able to sustain the illusion of two becoming One.

Notes and references

1 There actually was a headline with this title; see "This is your brain on God: spiritual experiences activate brain reward circuits," *University of Utah News*, November 29, 2016. Available at: https://unews.utah.edu/this-is-your-brain-on-god/.

2 As I note in the sections discussing each artwork, not every work lives equally up to this image of Affective Realism; they move in and out of the thematic, some living up to it and then not. I take as a basic premise that each artwork is multiple and can be interpreted in various, often contradictory, ways.

3 For images of Ralph Goings' artworks, see "Ralph Goings, realist painter," Ralph Goings. Available at: http://ralphlgoings.com/.

4 John Arthur, *Realists at Work: Studio Interviews and Working Methods of Ten Leading Contemporary Painters* (Ann Arbor: Watson-Guptill Publications Inc., 1983), 82.

5 For works by Chuck Close, see "Work." Available at: http://chuckclose.com/. For an image of April, see "April, 1990–1991," *The Broad*. Available at: www.thebroad. org/art/chuck-close/april.

6 See: Nikolas Carr, *The Shallows: What the Internet is Doing to Our Brains* (New York: W. W. Norton & Co., 2010).

7 Sherry Turkle, *Alone Together: Why We Expect More from Technology and Less from Each Other* (New York: Basic Books, 2011).

8 Adina L. Roskies, "Are neuroimages like photographs of the brain?" *Philosophy of Science* 74, no. 5 (2007): 861.

9 Sarah Knapton, "'Baby brain' really does exist, scientists say," *Telegraph*, May 7, 2014.

10 Ben Spencer, "Too much TV really does rot your brain," *Daily Mail*, December 3, 2016.

11 For discussion of "Neuro realism," see Eric Racine, Ofek Bar-Ilan, and Judy Illes, "fMRI in the public eye," *Nature Reviews Neuroscience* 6, no. 2 (2005): 159–164.

12 Leah Ceccarelli, *Shaping Science with Rhetoric: The Cases of Dobzhansky, Shrödinger, and Wilson* (Chicago: University of Chicago Press, 2001), 13–14.

13 See Suparna Choudhury and Jan Slaby, *Critical Neuroscience: A Handbook of the Social and Cultural Contexts of Neuroscience* (Hoboken: Blackwell Publishing, 2011); Eldritch

Priest, "Earworms, daydreams, and cognitive capitalism," *Theory, Culture & Society* 35, no. 1 (2018): 141–162.

14 See Cordelia Fine, *Delusions of Gender: How Our Minds, Society and Neurosexism Create Difference* (London: Icon Books, 2010); Cordelia Fine, "Is there neurosexism in functional neuroimaging investigations on sex differences?" *Neuroethics* 6, no. 2 (2013): 369–409.

15 Nancy D. Campbell, "Towards a critical neuroscience of 'addiction'," *BioSocieties* 5, no. 1 (2010): 89–104; Jordynn Jack, "'The extreme male brain?': Incrementum and the rhetorical gendering of Autism," *Disability Studies Quarterly* 31, no. 3 (2011), online. Available at: http://dsq-sds.org/article/view/1672/1599 (accessed August 1, 2017); Rebecca M. Jordan-Young, *Brain Storm: The Flaws in the Science of Sex Differences* (Cambridge, MA: Harvard University Press, 2010).

16 See Ivan M. Linforth, *The Arts of Orpheus* (New York: Arno Press, 1942); Jessie Poesch, "Ennius and Basino of Parma," *Journal of Warburg and Courtauld Institutes* 25, no. 1/2 (1961): 116–118.

17 Judy Illes, *Neuroethics: Anticipating the Future* (Oxford: Oxford University Press, 2017).

18 Kristina Fiore, "RSNA: violent video games may alter brain function," *Medpage Today*, November 28, 2011.

19 Raymond A. Belliotti, *Shakespeare and Philosophy: Lust, Love and Law* (Amsterdam: Rodopi, 2012), 116.

20 See Frederich Nietzsche, *Thus Spoke Zarathustra* (London: Penguin Classics, 1961), 41.

21 Eric Racine, Ofek Bar-Ilad, and Judy Illes, "fMRI," 159–164.

22 Chris Salter, *Alien Agency: Experimental Encounters with Art in the Making* (Cambridge, MA: MIT Press, 2015), 11.

23 Jenny Edbauer Rice, "The 'new' new: making a case for critical affect studies," *Quarterly Journal of Speech* 94, no. 2 (2008): 200–212.

24 Roberto Simanowski, *Data Love: The Seduction and Betrayal of Digital Technologies* (New York: Columbia University Press, 2016).

25 "Mind illuminated: mind and brain at the Mütter Museum," The Mütter Museum, June 4, 2015. Available at: http://muttermuseum.org/news/mind-illuminated-art-and-the-brain-at-the-m%C3%BCtter-museum/.

26 "Greg Dunn," *MedinArt*, 2013. Available at: www.medinart.eu/works/greg-dunn/ (accessed September, 2018).

27 Peter Galison, *Image and Logic: A Material Culture of Microphysics* (Chicago: University of Chicago Press, 1997), 22.

28 Deena Skolnick Weisberg, Jordan C. V. Taylor, and Emily J. Hopkins, "Deconstructing the seductive allure of neuroscience explanations," *Judgment and Decision Making* 10, no. 5 (2015): 429–441.

29 Victoria Woollaston, "A beautiful mind: Japanese-style art inspired by neuroscience reveals grey matter in much more colorful glory," *MailOnline*, December 15, 2014.

30 "What is sumi-e?" Giuseppe Signoritti website, *Sumi-e-it*. Available at: www.sumi-e.it/en/what-is-sumi-e/.

31 Robert Linsley, "The shape of colour [The Shape of Colour: Excursions in Colour Field Art, 1950-2005. Art Gallery of Ontario. Toronto]," *Canadian Art* 22, no. 3 (2005): 107.

32 Ibid., 108.

33 Martha J. Farah, Cayce J. Hook, and Gwendolyn M. Lawson, "Toward a reasoned approach to neuroeducation in an era of 'Neuroeverything'," Letter to the Editor, *Human Development*, May 8, 2014. Available at: https://neuroethics.upenn.edu/wp-content/uploads/2015/06/Pov-neurosci-worries-Letter1.pdf.

34 The quotations here describe Michel Foucault's concept of governmentality and derive from Tania Murray Li, "Governmentality," *Anthropologica* 49, no. 2 (2007): 275. For more, see: Michel Foucault, "Governmentality," in *The Foucault Effect:*

Studies in Governmentality, edited by Graham Burchell, Colin Gordon, and Peter Miller (Hempstead: Harvester Wheatsheaf, 1991), 87–104.

35 "SJR," *Scimago Journal and Country Rank*. Available at: http://www.scimagojr.com/journalrank.php?category=2801.

36 "Why is the BRAIN initiative needed?" *National Institutes of Health, US Department of Health and Human Services*. Available at: www.braininitiative.nih.gov/about/index.htm.

37 See "Brain Games," *National Geographic Television*; "Brain Bullet by Lee Benson," *Brain Bullet*. Available at: http://www.brainbullet.com/.

38 "Brain Bullet by Lee Benson," 1.

39 "The science behind NeuroNation," *NeuroNation*. Available at: https://sp.neurona tion.com/en/science/.

40 See Jenell Johnson and Melissa Littlefield, *The Neuroscientific Turn: Transdisciplinarity in the Age of the Brain* (East Lansing: Michigan State University Press, 2012).

41 For discussion, see Joseph Dumit, "The fragile unity of neuroscience," in *Neuroscience and Critique: Exploring the Limits of the Neurological Turn*, edited by Jan Vos and Ed Pluth (London: Routledge, 2015), 223–30, at 226; Ana Hedberg Olenina, "A case for neuohumanities," *Critical Inquiry*, September 21, 2017. Available at: https://critinq.wordpress.com/2017/09/21/a-case-for-neurohumanities/.

42 See Francesca Simion and Elisa Giorgio, "Face perception and processing in early infancy: inborn predispositions and developmental changes," *Frontiers in Psychology* 6, July (2015): 969. The comparison reference regarding mate preference is as follows: Scott Pitnick, Kate E. Jones, and Gerald Wilkinson, "Mating system and brain size in bats," *Proceedings. Biological sciences* 273, no. 1587 (2005): 719–724.

43 My opinion is that Davis' use of mirror neurons overlooks other theories and is overextended. Nevertheless, many thought it offered something. See Diane Davis, "Identification: Burke and Freud on who you are," *Rhetoric Society Quarterly* 38, no. 2 (2008): 123–147.

44 See Rocco Antonio Pisani, "Malcome Pines' contribution to group analysis," *Group Analysis* 43, no. 3 (2010): 328–336.

45 See Donald Hodges, "The neuroaesthetics of music," in *The Oxford Handbook of Music Psychology*, edited by Susan Hallam, Ian Cross, and Michael Thaut (Oxford: Oxford University Press, 2016).

46 See Vaughn Bell, "The marketing industry has started using neuroscience, but the results are more glitter than gold," *Guardian*, June 28, 2015; Grace Wong, "Brain scans gauge horror flick fear factor," *CNN*, September 29, 2009; Melanni Nanni, "Neural correlates of the natural observation of an emotionally loaded video," *PLoS One* 13, no. 6 (2018): e0198731.

47 Nora Rock, "How neuroscience awareness and evolutionary psychology can help lawyers avoid claims and offer better client service," *Avoid a Claim*, February 21, 2017. Available at: https://avoidaclaim.com/2017/how-neuroscience-awareness-and-evolutionary-psychology-can-help-lawyers-avoid-claims-and-offer-better-client-service/.

48 Ibid.

49 Greg Hickok, *The Myth of Mirror Neurons: The Real Neuroscience of Communication and Cognition* (New York: W. W. Norton & Co., 2014); David Gruber, "The extent of engagement, the means of invention: measuring debate about mirror neurons in the humanities and social sciences," *Journal of Science Communication* 15, no. 2 (2016): A01.

50 See Adrian Furnham, *Managing People in a Downturn* (London: Palgrave Macmillan, 2011), 122–124.

51 Tom Billington, "Educational inclusion and critical neuroscience: friends or foes?" *International Journal of Inclusive Education* 21, no. 8 (2016): 866–880.

52 See Joseph Dumit, *Picturing Personhood* (Princeton: Princeton University Press, 2004), 6–10.

53 See Jan De Vos and Ed Pluth, *Neuroscience and Critique: Exploring the Limits of the Neurological Turn* (New York, NY: Routledge, 2015).

54 Johnson Thornton, *Brain Culture: Neuroscience and Popular Media* (New Brunswick: Rutgers University Press, 2011), 1.

55 Bruce E. Wexler, *Brain and Culture* (Cambridge, MA: MIT Press, 2006).

56 Suparna Choudhury and Jan Slaby, *Critical Neuroscience: A Handbook of the Social and Cultural Contexts of Neuroscience* (London: Wiley-Blackwell, 2011).

57 Bahar Gholipour, "Giant artwork reflects the gorgeous complexity of the human brain," *Huffington Post*, June 25, 2016, para. 14.

58 The comment is made in reference to another work included in the *Mind Illuminated* exhibit but certainly holds true for "Brainbow Hippocampus." The phrase is also repeated in an interview with Dunn; see Linda Codega, "Elucidating the nature of consciousness through art: an interview with Greg Dunn," Garrison Institute, October, 20, 2017. Available at: www.garrisoninstitute.org/blog/human-consciousness-art/; Also see "Self reflected," Greg Dunn Available at: www.gregadunn.com/self-reflected/.

59 Andrea Baucon, "Leonardo da Vinci, the founding father of iconography," *Palaios* 25, no. 5/6 (2010): 361.

60 For discussion, see Liana De Girolami Cheney, "Leonardo da Vinci's Uffizi Annunciation: the Holy Spirit," *Artibus et Historiae* 32, no. 63 (2011): 39–53.

61 René Descartes, *Les Principes de la Philosophie. Introduction et notes*, translated by Guy Durandin (Paris, FR: Librairie Philosophique, 1970).

62 Kim Meeri, "Mutter Museum features artist's 'mind-blowing' images of the brain," *PhillyVoice*, July 7, 2015, para. 4.

63 "An illuminated mind offers Mind Illuminated," *PennMedicine News*, Fall 2015. Available at: www.pennmedicine.org/news/publications-and-special-projects/penn-medicine-magazine/archived-issues/2015/fall-2015/mind.

64 See Wolff who tackles this issue of the displacement of discourse: Janet Wolff, "After cultural theory: the power of images, the lure of immediacy," *Journal of Visual Culture* 11, no. 1 (2012): 4–7.

65 William E. Connolly, "The 'New Materialism' and the fragility of things," *Millennium* 41, no. 3 (2013): 401–402.

66 Ibid.

67 Samantha Coole and Diane Frost, *New Materialisms: Ontology, Agency, and Politics* (Durham, NC: Duke University Press, 2010), 2.

68 Wolff, "After," 4–7.

69 For discussion of the need to move beyond representation, see John Lynch, "Articulating scientific practice: Understanding Dean Hamer's 'Gay Gene' study as overlapping material, social and rhetorical registers," *Quarterly Journal of Speech* 95, no. 4 (2009): 435–456.

70 Connolly, "The 'New Materialism,'" 402.

71 The phrase is repeated by Marback and comes from Richard Marback, "Detroit and the closed fist: toward a theory of material rhetoric," *Rhetoric Review* 17, no. 1 (1998): 74–92.

72 See "Andy Kaufman's Tony Clifton 'Dinah!' incident (1979)," *Lost Media Archive*. Available at: http://lostmediaarchive.wikia.com/wiki/Andy_Kaufman%27s_Tony_Clifton_%22Dinah!%22_Incident_(1979).

73 Associated Press, "Las Vegas shooter's brain to undergo microscopic study," *Guardian*, October 29, 2017.

74 See Sherry Fink, "Las Vegas gunman's brain exam only deepens mystery of his actions," *New York Times*, February 9, 2018.

75 Paul Catley, "The future of neurolaw," *European Journal of Current Legal Issues* 22, no. 2 (2016), online. Available at: http://webjcli.org/article/view/487/651; Paul Catley and Lisa Claydon, "The use of neuroscientific evidence in the courtroom by

those accused of criminal offenses in England and Wales," *Journal of Law and the Biosciences* 2, no. 3 (2015): 510–549.

76 Tineke Broer and Martyn Pickersgill, "Targeting brains, producing responsibilities: the use of neuroscience within British social policy," *Social Science & Medicine* 132, May issue (2015): 54–61.

77 Warren Cornwall, "Did you knowingly commit a crime? Brain scans could tell," *Science*, May 13, 2017.

78 See: David Gruber, "Mirror neurons in a Group Analysis hall of mirrors: 'translation' as rhetorical approach to neurodisciplinary writing," *Technical Communication Quarterly* 23, no. 3 (2014): 207–226. For a popular example, see the CogniFit Blog: Andrea García Cerdán, "Mirror neurons: the most powerful learning tool," *CogniFit*, June 8, 2017. Available at: https://blog.cognifit.com/mirror-neurons/.

79 "Adolescent brain makes learning easier," *ScienceDaily*, December 17, 2017

80 Bruce Redwine, "The uses of memento mori in Vanity Fair," *Studies in English Literature, 1500-1900* 17, no. 4 (1977): 658.

81 Marjorie Garber, "'Remember Me': 'Memento Mori' figures in Shakespeare's plays," *Renaissance Drama* 12 (1981): 4.

82 For an overview, see Jane Prophet, "Neuro Memento Mori," *Jane Prophet*. Available at: www.janeprophet.com/?p=326; "Neuro Me-mento Mori," *NeuroMementoMori*. Available at: www.neuro-memento-mori.com/; Jane Prophet, "(Projection) mapping the brain: a critical cartographic approach to the artist's use of fMRI to study the contemplation of death," School of Creative Media, City University of Hong Kong, 2015.

83 Redwine, "The uses," 658.

84 See "Neuro Memento Mori, projection mapping and documentary video," *Jane Prophet*, June 3, 2015. Available at: https://vimeo.com/janeprophet.

85 Ibid.

86 Ibid.

87 Ibid.

88 Benjamin De Mott, "Science and the rejection of Realism in art," *Synthese* 14, no. 5 (1963): 390.

89 Slavoj Zizek, *For They Know Not What They Do: Enjoyment as a Political Factor* (London: Verso Books, 1991), 241.

90 Ibid.

91 "Go your own way" is famously a song composed by Fleetwood Mac, produced by Ken Caillat, Richard Dashut, and Fleetwood Mac on the 1977 album "Rumors" released by Reprise Records, Burbank, CA.

92 Laura Jade, "Brainlight," *Laura Jade*, para. 1. Available at: http://laurajade.com.au/interactive-brain-light-research-project/.

93 Laura Jade, "Interactive Brainlight: brain sculpture lights up your thoughts," YouTube.com, October 9, 2015. Available at: www.youtube.com/watch?v=ZQmA5X47ewU.

94 Jade, "Brainlight," para. 1.

95 Louis Althusser, "Ideology and ideological state apparatuses (notes towards an investigation)," in *Lenin and Philosophy and Other Essays* (New York: Verso, 1970), 11.

96 Here I follow Brian Massumi's general definition, treating affect as undefined disposition of feeling and emotion as a culturally defined, internally recognizable feeling state. See Brian Massumi, *Parables for the Virtual: Movement, Affect, Sensation* (Durham, NC: Duke University Press, 2002).

97 For discussion of a Will to Freedom, see Stephanie M. Batters, "Care of the self and the will to freedom: Michel Foucault, critique and ethics," Senior Honors Project, University of Rhode Island. Also see: Michel Foucault, "What is critique?" in *The Essential Foucault* (New York: The New Press, 1997), 263–278. For source material on a Will to Power, see Friedrich Nietzsche, *The Will to Power*, translated by Walter

Kauffman and R. J. Hollingdale. Vintage Books Edition (New York: Random House, 1968).

98 Nietzsche, *The Will*, 11.

99 Batters, "Care," 1–2.

100 Jade, "Interactive," video.

101 At present, I have not seen the work used in this way, but I hope it will be expanded and explored for other potentials.

102 Ibid.

103 See *The Standard Edition of the Complete Psychological Works of Sigmund Freud (S.E.)*, Vols. 7 and 9, translated by James Strachey and Anna Freud (London: Hogarth Press, 1886–1889).

104 Monica Michlin, "Open your eyes wider: overexposure in contemporary American film and TV series," *Silages Critiques* 17, Part II (2014): 1–3.

105 Constantine Sandis, "Hitchcock's conscious use of Freud's subconscious," *European Journal of Psychology* 5, no. 3 (2009): 56–81.

106 Laura Mulvey, "Visual pleasure and narrative cinema," in *Film Theory and Criticism: Introductory Readings*, edited by Leo Braudy and Marshall Cohen (New York: Oxford University Press, 1999), 836.

107 See Alastair Sooke, "Lucian Freud and the art of the full-figured nude," *BBC Culture*, May 12, 2015.

108 Donna Haraway, "Situated knowledges: the science question in feminism and the privledge of partial perspective," *Feminist Studies* 14, no. 3 (1988): 581.

109 Ibid.

110 "Genesis 3:7," *Berean Study Bible*.

111 Ibid.

112 Bruce Boucher, "Sinners and saints," *New York Times*, August 8, 1999, para. 5. Available at: https://archive.nytimes.com/www.nytimes.com/books/99/08/08/reviews/990808.08bouchet.html.

113 Francesca Biliani, "Return to order as return to realism in two Italian elite literary magazines of the 1920s and 1930s: La Ronda and Orpheus," *The Modern Language Review* 108, no. 3 (2013): 845–850.

114 Brent J. Piepergerdes, "Re-envisioning the nation: Film neorealism and the post-war Italian condition," *ACME: An International E-Journal for Critical Geographies* 6, no. 2 (2007): 231–233.

115 Marcus Millicent, "Freccerò on Blow-Up: toward a macro-vision of Italian film," *MLN* 124, no. 5 (2009): S225.

116 Ibid.

117 Ibid., S226–S227.

118 Nietzsche, *The Will*, 3–4.

119 Werner Hamacher, "Guilt history: Benjamin's sketch 'Capitalism as religion'," translated by Kirk Wetters, *Diacritics* 32, no. 3/4 (2002): 81–106.

120 See "(I Can't Get No) Satisfaction," *The Rolling Stones*, music composed by Mick Jagger and Keith Richards, 1964.

121 Bruno Latour, "On some of the affects of capitalism," lecture given at the Royal Academy, Copenhagen, February 26, 2014, p. 1. Available at: www.bruno-latour.fr/sites/default/files/136-AFFECTS-OF-K-COPENHAGUE.pdf.

122 Ibid.

123 Ibid.

124 See 2 Corinthians 4:18, The Berean Study Bible. Available at: https://biblehub.com/bsb/2_corinthians/4.htm.

125 It is probably fair to say that working to see the face of God, or to see more compassion come into the world is that work referenced and hoped for, as much as salvation in the traditional sense of entrance to heaven.

126 For the story of the women, see Exodus 1, The Berean Study Bible. For the discussion of God making Himself more fully known to Moses, see Exodus 6:3. Available at: https://biblehub.com/topical/exodus/1-1.htm.

127 I make this claim on the premise that Nietzsche wanted us to understand a world as a Will to Power, of forces, and grasping that insight was, itself, a means to empowerment. That is, there is nothing mystical about Nietzsche's order, and the emotional, although pertinent, must be subdued, it seems to me, to achieve a transcendence, which is primarily a mental overcoming made and then forced onto the body feeling a need for it.

128 Heidegger's position is well explained in Martin Heidegger's "Nietzsche's Word: 'God is Dead'" essay, which appears in: Martin Heidegger, *Off the Beaten Track*, edited by Julian Young and Kenneth Haynes (Cambridge: Cambridge University Press, 2002), 157–199.

129 Julia Kristeva, *Approaching Abjection, Powers of Horror* (New York: Columbia University Press, 1982), 1.

130 Ibid., 2.

4

BRAIN ART ON THE ONTOLOGICAL STAGE

Up to this point, I have turned to an eclectic group of resources—ranging from the Critical Neurosciences to Continental Philosophy—to interrogate the symbolic and material intricacies of six acclaimed contemporary brain artworks. What has emerged for consideration are two themes: neurosensuality, or the sensual and intimate relations of neuroscience to everyday life, and Affective Realism, or the technical manufacturing of an illusion of greater access to and more real understandings of confounding experiences through the use of brain imaging. Detailing these, I have tried to express their affordances and limitations. I have tried to be fair in my critiques, arguing that brain art can both communicate a beautific neuroscientific future and a troubling one at the same time. A work can propose the medical value of non-invasive imagery into the body and still promote new forms of surveillance. It can aggrandize "Nature" while implying cultural constructedness and highlighting the nuances of experimental selections. It can swim in the pool of scientism yet, with one small gesture, call the objective outside into question and return the burden of proof to the neurosciences. This is, after all, why we love art. It inspires and challenges, asks us to question our most basic commitments, and leaves us suspended at just the right moment and in just enough empty air to feel the world ever full of wonder.

But more than celebrate the contravening soul of art in playing both sides, I hope to also show that neurosensuality and Affective Realism are not necessarily negative themes. Yes, they have consequences, but they point to how we are imagining a brain and what we want to do with a brain. My suggestion is, ultimately, that we walk away not with an image of the fleshy brain but a representation of its broader socio-cultural articulation, i.e. neurosensuality and Affective Realism are embedded in The Brain, and they not only become apparent in brain art but unravel a bit when subject to creative material expressions. "This is not a pipe," Magritte playfully reminds us.[1] The Brain, like the pipe, becomes sets of colors. And then the colors can be seen as paint. And then the paint becomes molecules and light waves. And then we refocus the eyes and step back to question what the hell Magritte chose a pipe for anyhow? We can conjure associations with old men in the Paris elite smoking together and telling each how, exactly, the world *is* in every respect. And then we laugh. "This is not a pipe" is a representation seen anew.

So the question now: is brain art achieving what Magritte achieved, or is it largely stuck in a strict representationalist orientation fitting to neuroscience's methods and correlations? Where else can brain art go? What more can it do?

Contemporary brain art is not necessarily unified as a movement even if it leans collectively toward the representational and the idealized. The works explored in past chapters seem to have some commonalities, however. They prioritize beauty, highlight wonder, and respectfully present the neurosciences to general audiences. Challenges to the power and politics of the institutional, disciplined apparatus are hidden in the weave. A further observation can be made: brain art, as I see it here, revolves largely around objects prepped for repeated installation. Although some brain art might diverge from this formula and be performative—such as *A-Me* or *Interactive Brainlight*—they are tightly circumscribed. They do not allow for much evolution of the core concept. Alterations in the performance are minimal; audiences cannot impact the work's structural form nor interrupt what the artist hopes to achieve when presenting what is recognizable as "neuro" or as brainy. Participants of *Interactive Brainlight* might be able to alter the colors of the brain sculpture, for example, but the thing remains a brain, enabling a repeat performance such that the sculpture can accommodate various venues and function more like a traveling rock show. The same is true for *A-Me*. These are not open-ended creatures shifting their material or emotional directions to craft unique life cycles even if they are participatory in some respects.

A contrasting idea of art is therefore worth consideration. Pickering's art of "ontological theatre" is a protagonist working on behalf of the audience and against the inscribed lines of the representational.[2] The art of ontological theater is not geared toward a static object on the wall for the half-interested viewer holding a glass of wine and sniffing out business contacts. It is not the better object for understanding an object nor the magnificent signification of a Thing's deeper import. Art of ontological theater demands the moment, unfettered, opening itself in a passing engagement to the unexpected impositions of an outside, changing with every installation, generating an experience beyond the control of the artist.

Art in this mode is probably best detailed in relation to Deleuze's rejection of discrete things in themselves, as somehow independent, self-contained, and stable. Drawing inspiration from the Baroque style of "low and curved stairs … spongy, cavernous shapes,"[3] Deleuze presents all things as material foldings and unfoldings, favoring fluctuation, "a continuous variation of matter."[4] He pushes artists and philosophers to embrace an emergent and often disconnected order wherein a transformation, word, or machine is not "prefigured by those units in the course of logical development."[5] Ultimately, this stance compels Deleuze to consider alternative possibilities for art—as something that might exist, as he says, "in extension."[6] He imagines a sculpture extending into its base, into the floor, into the landscape, into the city, with "each art tending to be prolonged and even to be prolonged into the next art," a "wrapping and unwrapping," a performance of folds.[7] Art as ontological theater pursues this Deleuzian concept.

But in Pickering's view, art of ontological theater must also stress experimentation with the affordances of a local environment. Doing so, he argues, reflects the work of early cyberneticists who see living organisms as always having an opportunity to develop because they are not independent of environments but "adaptive systems."[8] Ontological theater, then, is "ontological" precisely because it de-laces a definitive answer about the way that the brain (and larger world) forever is or will be. It meshes or embeds the stuff presumed exterior—of the world—in whatever is put forward as a Thing. The theatrical manifestation thus asks how Things might change in a strange or psychedelic way if pushed into odd encounters and over our imagined limits.

Artworks harboring no claim to objective knowledge, happily swimming in contingency, seek outlandish new versions of themselves. Dedicated to emergence, the art of ontological theater dances amid the potentials of relationality without the desire to have a say in reassembling structures. It embraces a multitude of agentive forces without presuming that networked lineages are available for the artist, or any individual, to find on one's own. It celebrates and relies totally on the unexpected. Accordingly, what aspects of Things might escape us, or whether absolutely everything approaches other Things in the same way or not, matters little in contrast to the action on the stage. Whether ontologies are flat or unmanageably shifting in various hierarchies—how Things *really are*—is subordinated to the right now that exposes *what lives and what happens*.

Cut against ontological theater, some of the domineering and objectifying works of brain art look like big ugly Modernist set-pieces imbued with Cartesian presumptions. Any interdisciplinary cheerleading for new techno-artistic practices dissipates in a soup of hard-nosed scientism in too many works. Other artworks, most of those that I have focused on in this book, better wet the palette, retaining dimensionality and unsettling depth. However, with that said, it may not be a mistake that many works of brain art, themselves built from the tools of the neurosciences, are not easily classified as non-Modern. Neuroscience is disciplined and regimented to generate circumscribed studies of the nervous system able to be replicated in a laboratory and, as such, an art that adopts those methods and tries to find creative avenues for those ways of seeing often repeat the equation. In fact, sometimes we might even suspect that brain art is a neuroscientific publicity piece. Far from ontological theater, we sometimes discover in brain art a theater of the absurd—so resolute and serious is a brain-based work sometimes that the critic doubts whether anyone in the museum is aware of how the work itself, in being positioned as "a brain" or in being realized through strange materials (silk or paper or paint or LED lights), undercuts its claim and its status. In those cases, we must then question our own reading. Turn ourselves back around. See the puppet performance for what it is. But even when we do, we cannot help but think: there is a seriousness about the "actual" here, something purported about the "real" human, isn't there? And we wonder: can brain art be more than stabilization and extend beyond a dedication to the technology, not position human phenomenal experience as discrete or capturable?

In this chapter, I consider what brain art can do to escape the impulses of visual objectification, universalization, simplification, rigidity, and control. To be less serious, to play up the jokey opposition inherent in the serious enterprise, to break the third wall, to allow the clown in the audience to paint a big white brush stroke over the realist landscape that seems perfect in every detail. Aiming for positive contribution, I try out a few examples, first focusing on playful works that draw on postmodernism and the sardonic trends of the mid and late 20th century. These circumvent any simple critique by openly displaying contradictory and confounding impulses. Charlotte Rae's "Marilyn's Brain," as one example to be discussed, is more than simplistic commentary on popular neuroscience magazines, just as Andy Warhol's "Marilyn Diptych" is more than a play of colors signifying celebrity glam.

Unexpected associations force ruptures, but more than knocking over significations, these ruptures infect the body—they produce a-signifying reactions, sickly doubts in the gut about our ways of being and knowing. In those moments, the body bubbles up resistance to the mind's obsession with predictability, organization, and simplicity. The most effective forms of neurosensuality and Affective Realism in brain art turn against themselves. They plead with us to reconsider what we know—and do so by exposing the sudden surprise of the moment, the now, the nuance of alteration in the alternative whiff of air, the sewing needle stuck into the center of the brain scan during the opening of the exhibit, the sharp squeal of excitement from two rows back when we expected reverential tones.

But playful artworks, in merely being playful, do not necessarily take the full dive into the non-Modern. So we consider performative art. Works such as *The Brain Piece* by Jody Eberfelder encourage us to throw the intricately mapped brain overboard and stand on the edge to see if it drowns.[9] That work will be discussed in this chapter. But the point is: when we re-see the brain cybernetically, as Pickering does, as adaptive and performative in real-time and living in concert with the dance of the environmental milieu, then we wonder where the brain might be. Can it be so discrete and differentiated? Through creative experimentation, or art of ontological theater, we attune to the vibrancy of materiality and interrogate how we feel about ourselves as brains and then rethink ourselves as bodies. We risk feeling through our arms and legs and stomachs, puking all over the floor when environments think for us. And we risk admitting that the suasive means are not three good, logical reasons but the smell of warm bread, the kick of a cup of coffee, or the lowering of the lights and a good martini.

In reimagining brain art, we reimagine the limits of the brain. We try to discover if there is freedom in the instantaneous life. What would it mean to live without a materialist core, a guide locatable like a brain? Ultimately, we step back and wonder if dedication to inventive possibilities in ecological and continual Becoming is a pipe dream (another Magritte painting, but one saying "which pipe?"). Is ontological theater a gasp of erotic pleasure that dissipates like a wild night of uninhibited sex? If we dance with the rhythms of the forest like a wild wolf having a dream, will we wake up suddenly feeling out of place, worthless, and not know what to do with ourselves?

Scan 13: The Brain of our ontologies is a Jekyll and Hyde

New Materialism is a broad intellectual movement pursuing complex integrations of history, biology, and symbolicity.[10] The New Materialist analysis of human conditions becomes an issue of thinking through various compatibilities among Things, i.e. giving attention to the "intra-action" of co-forming beings and their unique coming-togethers, as Karen Barad says.[11] Or, for a slightly different take on the same theme, New Materialisms can be said to focus on ecologies of "ambient environs" and their affectabilities, as Thomas Rickert advocates.[12] The general idea is to favor complexity and the fluidity of Becoming. The scholarship seeks greater nuance when asking, for example, how doctors identify a disease state. The knowing-how (such as looking through a microscope) makes a perceptible difference to determining the existence of a knowing-what or the kind of thing (such as a type of disease). The means of access is deemed so important that New Materialists pursue methodological expansion and pluralism.[13] Proponents, as a result, try to work in groups or find different ways to study a phenomenon; analysis of texts, on-site observations, experiments, etc. help researchers to think differently, to look with and alongside the things that they study. The guiding idea is simply that intervening environments alter what is going to be (and able to be) made and seen.

This drive to be more interdisciplinary and have a wider view fits social conditions where work is increasingly mediated and globalized. Charting matter as relational (about the fit or the shape of things engaging) advances a politics that extends a principle of shared living and cooperation out to all things. When transoceanic exchanges and affective connections happen in an instant, it becomes useful to think the world, writ large, in the same way. Functionally, New Materialist thinkers, such as Barad and Rickert, aim not to pretend in any way that we do or can stand independent and free.

So I want to assert that the brain can be reformulated in this way as well. Not as a Thing, isolated nor as a Thing pushing into other discrete Things but, rather, as a constant generative. We cannot pretend through the generative brain and its agentive linkages that we are independent and free.

If understood as Becoming, as multiple, and as eventful, the brain disperses. At that point, saying that the brain has "braininess" seems strange. Braininess would not be of the brain in any easy or direct way.[14] This assertion, too, is a blending of the social and the ontological at a moment when global reconfigurations compel good questioning about the viability of static concepts of Being.

Studying a brain as "intra-actional," i.e. drawing from Karen Barad, means upending science's purported control over external variables and remaking, over and over, the disciplined and rigid apparatus of experimentation to see what else can come into appearance. Yet, as might be intuited already by the reader, there are drawbacks and benefits to pressing to the issue. First, the benefits.

Pursuing an instantaneous brain that is intra-actional makes the abstracted The Brain more difficult to hold in place. This is good for neuroscientists because The Brain is an extreme overestimation of what neuroscience is able to actually see, do, and say. Likewise, it is good for social relations in so far as discussions about "normality" as well as human exceptionalism amongst the animals moves into the background. When the formation of the entity (the organ) is shifting and emerging amid environmental landscapes, it is difficult to appeal to the isolated Thing to craft an unassailable reality for human life. An intra-actional brain opens up space to think the thought: "there is no brain."[15] If local environments and co-forming beings perform existence, then hyper-focusing on a brain to answer questions about subjective experience is pursuing a distraction—like obsessing over exhaust trails when trying to understand what a car is or, more broadly, a world of traffic.

However, there are also drawbacks to foregrounding a messy mix of instantaneous emergences. A destabilized, intra-actional brain must contend with neuroscientific studies showing "universal circuits,"[16] "common valuation scales" within economical brain systems,[17] and similar "modes of processing" for experience despite local variables.[18] The Brain, as the grand abstraction, takes shape from popular and disciplinary integrations of various results, and this seems to limit the power of hyper-local environmental coming-togethers to overrun neuroscience with a bevy of brainy iterations. As Rose puts it when driving home science's dismissal of cultural theory:

> Who needs vitalism when the complexity of living systems can be broken down into describable interactions between specific kinds of parts, their living processes can be reverse engineered, the parts and their properties can be freed from their origins in any specific organism, and reassembled, first in thought, then in reality, to produce whatever outcome you can dream up.[19]

In other words, who needs uniqueness and particularity when you can rely repeatedly on commonality and systematized generalizations? Who needs an analysis of "ambient environs" when the organism is so predictable?

Yet, the invigorated environmental, indeed compassionate and ethical, value embedded in New Materialisms attempts to discover how we can live and develop differently. In that sense, New Materialisms escape a framework for living based on what Rose calls the reliable "parts" and "properties." Too much dependence on technocratic modularity endangers the variability of human expression. But more so, a scientific angle should also recognize how representationalism shutters off emergent materialities.

The Jekyll and Hyde of New Materialism appears now: a vibrant relational ontology dedicates to co-being and emergence, leaving scholars wanting to tell a story about matter's instantaneous expression while yet still claiming to be absolutely certain about the principles of matter. The dedication to grouping all of matter together, treating it as essentially the same—relational—shows that the drive to pluralism does not necessarily extend to ontological essentials. These ontologies, too, have some unshakable first realities. Ontological instability is only relational or aesthetic, without those concepts being subject to internal shakability. The unyielding importance of a foundational "real" sets before us the connubial relation we have to material simplicity.

There is a cognitive dissonance there—a gap between the just now and the always of coming into being. I read this as a kind of Jekyll and a Hyde. Although generativity and complexity appeal to the scholar's desire to grasp the unknown while recognizing the inability to see any totality, these concepts simultaneously forward (even as they rebuke) how all Things definitively function. Thus, complexification must deny its own superiority in the ontological imaginary privileging it. Generativity, as a kind of engine of complexification, once positioned as a master trope, rests on the singular guiding concept that all matter is and has always been One kind of way (relational) despite how wild it seems.

To be fair, the reader might balk at any critique of having a One for matter. Admittedly, proposing no foundation—an undefined everything—does little good. What can we do with that? Really? Besides, we all breathe the composition of earth's air, walk carefully through the ice and snow, and feel the equations of gravity. We sense, fairly often, that we want more of that. Rules. Order. Predictability. And, admittedly, there are good reasons for matter to be conceptually unified. The world is presumably stable precisely because of some basic coherence across the formations of matter. We don't go flying out into space. Not yet anyway. We, thus, seek out overarching principles. As Jogelakar notes, "The search for unification is in one sense a search for harmony, a desire to view the whole universe through the lens of a single elegant law," and because others have unified past theories, such as Faraday and Maxwell who unified electricity and magnetism, the idea that all Things do work the same way and derive their behavior from the same set of laws is plausible, even if inexhaustible diversity is closer to the inner life of Things.[20] These contentions offer good reasons to dedicate oneself to a somewhat predictable and precise ontological description, i.e. to allow Things to be fluid but make the principle of fluidity, at least in theory, unwavering.

Now we must reverse course. Continual dissatisfaction and failure, like material loss and absence, disrupt ontological assurances. Destroying the coherence of matter, in fact, turns the body fully around to re-embrace that terrible feeling that an open-ended philosophy gives—of being stranded. In fact, reducing experience and decision-making down to a brain likely increases the *felt sense* that ontological principles cannot be resolved scientifically. That is to say, firm ontological presumptions do not necessarily eradicate affects. Stepping into a bottomless depth of materiality where dimensional shape-shifting can render every curtain of "the real" pursues another, entirely different and more magical vision of the universe, one which cooperates with a body that does not know itself.

Of course, I cannot but presume an ontology in discussing neuroscience and brain art. But I actively pull back from appeasing the desire to concretize any choreographed proposal. The suggestion, then, is that One—and yours as well—must be always again questioned and dismantled even if it must be embraced to think the next thought. We must risk letting go, ever again, to feel through what it might be like to let go. This is to advocate endless rethinking and exploration.

Upturning The Brain to rethink a brain as radical relationality means shoving off again. The Brain must come untethered from the rope and then we can allow the tide to drag the body back out to sea. There is no point to being stranded on a beachhead if it cannot satisfy. The pirate rum buried there will only last so long.

Postmodern play with neuroscience

The Modernist adoption of flat visual surfaces turns attention to form and materials —painting as painting without the pretense of realism.[21] In postmodernism, however, flatness turns on itself, adopting a new, playful, or cynical recognition of constructedness. Joselit puts it this way: "the psychological depth [of Modernism] undergoes deflation, resulting in a [postmodern] visuality in which identity manifests itself as a culturally conditioned play of stereotype."[22] As in Andy Warhol's "Marilyn Diptych," the deployment of mass production techniques, flat repetitive color blocks, and celebrity adulation exemplify both desire and disgust for objectification and stereotype.[23] The true or locatable Self dissipates in a (Campbell's) soup of intertextuality. Here, we run square into Charlotte Rae's work titled "Marilyn's Brain," a flat, glossy image supposedly showing Marilyn Monroe's brain, drawing explicitly from Warhol's work.[24]

Playful blocks of color reminiscent of Warhol's pop images call attention to neuroscientific celebrity and consumer obsession. Brain imagery is part and parcel of consumer culture's wild self-spectaclism, the work seems to say. The intertextuality suggests that neuroscience pursues hip topic areas and is embedded into sellable pop-psychological contexts. The exposé of "Marilyn's Brain" suggests another radical extension of Capitalism—science gone wild with private configurations sold to the highest bidder. Yet the work displays the limits of the medium and the brain scan's inability to provide insights about Marilyn, i.e. her real experience, her deeper sense of identity. Or, "Marilyn's Brain" might be

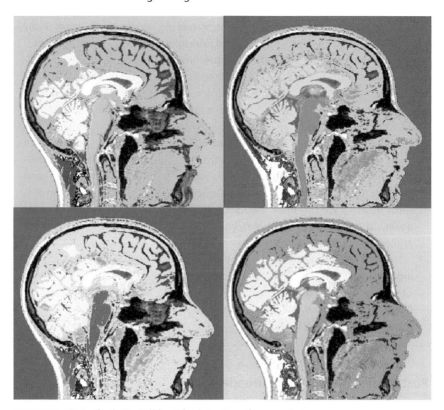

FIGURE 10 "Marilyn's Brain" by Charlotte Rae, by permission.

read as intensifying the male gaze by moving right beyond the exterior surface to penetrate her body, insinuating that neuroscience is yet another arm of masculine domination.[25]

Overall, Rae's work stresses the value of ironical social commentary with respect to the heightened epistemological status of the neurosciences. Like Warhol's works, "Marilyn's Brain" insists that "we rethink the meanings of consumption, collecting, visibility, celebrity, sexuality, identity and self-hood" in an age of brain scans when the potential for monitoring private thoughts and regulating others based on neural imagery grows more real everyday.[26] Here, "Marilyn's Brain" remakes Warhol's salacious social interventions in context of the contemporary brain sciences; Rae adds a layer of audacity to Warhol's infatuation with what is sexy and sensational amid the contemporary cult of commodification that he so masterfully parodied— and the work does so by applying neuroscience to the function of the paparazzi, peddling the image of the celebrity brain.[27]

Looking at "Marilyn's Brain," one finds it difficult not to see neuroscience as intertwined with the popular realms that contribute to its fame. It is difficult not to develop critical interpretations of neuroscience upon seeing the work. Although in some ways a Modern piece—overtly representational and hyper-focused on the effect of the colors and the surface—the work is also drawing upon the post-modern in flattening out any claim to "the real" within the play of circus style consumption and pop-bang fragmentation. Who knows what Marilyn's brain is or why it is colored red, blue, yellow, and pink? Maybe it is because she's so girly or so fiery or so rebellious or so beautiful? Maybe it is because she is so glamorized by consumerism? Maybe the brain, somehow, reflects her social identity? Does her brain become inscribed with social interpellations and end up reflecting an altered materiality in accordance with a new social real? The artwork, "Marilyn's Brain," certainly does. But no one is going to take this work as an authentic neuroscientific inscription. Consequently, the artwork, as an iteration of brain art—i.e. as a creative depiction of the brain—turns away from scientific realism, transforming into a commentary on the trajectory of art as well as the social productions of the neurosciences.

We might call "Marilyn's Brain" a critical neuroart, just as we might call War-hol's "Marilyn Diptych" a critique of power that shows us how consumerism absorbs all realms, visible and invisible. In like manner, Warhol's piece implicates art as a market, showing art's intrinsic affectation with and development through fame afforded by its ties to capital. Rae's version, then, would suggest the same for the neurosciences. It becomes an indictment of the financial motivations that drive excitement about the neurosciences and serve as background for too many studies of brains.

Warhol's artwork can also be understood as iconographic, which is the way that Kristeva positions it in her 1998 Louvre exhibition. There, Kristeva sees "Marilyn Diptych" as a suppression of "thought and death represented by the head" manifest-ing as iconography that intends to live forever. The head not only reminds of the eternal divine, but the female head also imparts comfort.[28] As a result, Warhol's "Marilyn Diptych," Kristeva argues, offers the calming face of the woman—histor-ically and psychologically a mother who gazes at her children—to soothe bodily feelings of separation and yet make us aware of our own inevitable death through the sight of her face, which is older and aging. Marilyn's face, an eternal beauty, thus, enacts a repression of what she herself can represent as an artwork.

"Marilyn's Brain," however, exchanges Marilyn's face for the neuroscientific image of the interior. In this exchange, the role of iconography as that which "transfers the invisible to the visible in the fixed economy of icons" positions Mari-lyn Monroe as icon of the invisible Self, or the looming death of the Self.[29] She becomes an icon of a search for what is real. An icon of desire in the absence of any substance. An icon of emptiness. An icon that is not merely the sexual desire directed toward the Hollywood appeal of the exterior but also an icon of the inter-ior, or the appeal of wanting more and simultaneously wanting to expose it all.

In the history of Modern art, there is a definable strain of works focusing on representations of the artist's own internal states. Such works themselves repre-sent the role of art to unveil what is hidden; and the best of these do so through the practices used to construct it. Think of Jackson Pollock's "dense web of lines and forms, like an unfathomable galaxy of stars," which calls attention to an inherently active and uncontrollable Being not easily captured by old Freudian categories and does so through dripping and spilling paint.[30] In achieving the push beyond the usual through the invention of a new painting technique, Pollock's works resuscitate the artist-genius having special insight even as the wildness of the canvas—the visual shock of the new—emboldens the old narrative of the artist as tortured by society's normalities. Pollock's works are, accordingly, said to be "dripping with genius,"[31] and the word "'Genius' recurs like a pious prayer" whenever his works are viewed.[32] If he reveals what is hidden, then he probably does not reveal his own psychology as much as the underlying discourses that structure how we see him and art.

The shift to postmodern art's opulent stylizations does not mean it cannot fulfill this same role, namely, to make visual hidden realities and underlying stories about art and artists. However, a gloating emphasis on the sell-ability of internal exposés yanks the seriousness of representations out onto Broadway and down toward Wall Street. Postmodern artworks tend to suggest that the power to define human Being, far from being uncontrollable, is knee-deep in economic and structural conditions; what is uncontrollable is only the need to creatively make fun of ourselves when we accept such conditions. For Warhol, as perhaps for Rae, a kind of hyper-performance of those conditions seems irresistible. And useful. Indeed, critical postmodern self-awareness finds playful clowning around to be a handy escape from any judgment that painting glamor and selling self-recognition to the highest bidder is old fashioned and still Modern.

Despite not confessing an authentic or unique Self—quite the contrary!—Warhol is, nevertheless, and ironically, marked as a "genius" because of his ability to unveil something hidden and perform it through his art practices. Still, his attunements to society's Capitalist irrationalities and his boldness to perform them like royalty, right out in Manhattan's hottest clubs, makes him a delightful postmodernist. If we call him a "genius," then we do so with a self-aware criticality (hopefully). He paves the way for Jeff Koons and others to produce Wal-Mart sized art with the same basic impulses, heightening the scale of all associated effects as if to say there is no roof anymore for consumeristic pleasures.[33] Maybe the next turn needs, then, to be back to the inside of the body, the most intimate of spaces—an exploration of the most literal kind, which would also be irresistibly cheeky or sardonic.

When considering "Marilyn's Brain" in juxtaposition to Warhol and Pollock, we note how the work expresses both Modern and postmodern impulses. On the one hand, the piece suggests that we, as audiences, have not yet stepped out of a Modern visuality even while loving postmodern sarcasm. We are stuck in a moment, a limbo of fascination and confusion. We want it all. Firm ground. Real genius. Consumeristic self-awareness. A locatable Self. We want to sell ourselves back to ourselves and have fun doing it. But just like the old days before the muddle of Derrida and the like, we want it all to be telling us what is really, truly real. Or else, or else, or else, it cannot be worth very much, can it?

From this stance, we consider neuroscience. We ask ourselves: what are we doing socially with these radical, invasive images of the interior? What are our intentions? Does this severed head, a decapitation, not tempt us to associate death with neuroscience? Does the genius want to kill off what is individualized to the artist-as-creator in recycling the technical reproduction as that which forges a more innovative view of the interior?

Perhaps the Hollywood brain up on the museum wall is another John the Baptist. Marilyn Monroe's severed head positions her as a 20th-century prophet of a new Christ on the verge of breaking another social and interpersonal veil. She comes in the image of Marilyn, a Holy Mother of Fashion Week. As a goddess of Hollywood, she's the forerunner of The Brain's strong link among media, the body, and economic viability. Her face, iconic in its allure and representative of her status, reminds of her many lovers as much as her indignities. The image foretells, we realize now, her eventual death in the bathtub. Her brain, suddenly able to be the complicated psychological subject par excellence, hides secrets even as it is reduced down to palatable movie-land significations—a dress, a smile, a figure, a witty line, and a wink. We might go so far as to say that her brain scan, like a John the Baptist, "prefigures the peace promised by a new religion" in presenting a decapitation as the unavoidable pathway to something greater yet to come—greater knowledge.[34] In this reading, "Marilyn's Brain" acts as the prophet of a neuroscientific age of brain scanning promising to use visuality to save us from our own obscurity while yet foreshadowing the need to commit violent segmentations of the body to achieve the vision.

We now entertain an opposite thought: we wonder if selling Marilyn's brain scan is a commentary on the death of a form of science that adamantly adheres to a Modernist concept of a world tied to the Enlightenment. In other words, if the repetitive color blocks of Marilyn's brain indicate a neuroscience that has turned decidedly social and "artsy," then the image separates science from its usual claim to objective knowledge production and the mechanical automaticity of images of the body. In this case, the image does not promise to save us but acts more like Marilyn herself: it gives us a wink and a smile, telling us *it is all just a bit of good fun*. Perhaps it also foreshadows the death of art's Enlightenment moment in the way that it does not pretend that the artist is an innovative creator of the representation; the agency goes almost entirely to the technical process. The image could, of course, also call both art and science into question—two formations whose disciplinary lives are not independent and whose geniuses must come to see each other through the other's machines.

Then there is Kristeva still hovering, waiting for her right treatment. She insists that Marilyn pictured as a female head represents an anxiety induced by "the castrated woman's sex."[35] In this reading, the image lures us in mascara colors to face a harsh reality—one where masculine power seeks to maintain itself through Othering and where it dehumanizes through violence. Objectifying Marilyn as a floating head means not only avoiding her body, but separating her from any strength, treating her as an object open to any segmentation and, simultaneously, avoiding a feeling of loss, a feeling Kristeva might well argue is already operative in the masculine Self-concept that so fears castration.

Further, if the severed head points generally to a historical "negotiation with fear of death" and the separation of Mother from child, or "the loss of the maternal face," as Kristeva suggests,[36] then the brain of the iconic woman must intensify these meanings. Being neuroscientific in style, the meanings must then be applied to scientific visuality. Indeed, what can we make of the world's most desirable and sellable woman pictured as a brain? Suddenly, "Marilyn's Brain" is not only about the trajectories of brain art but about neuroscience; the work announces a risk: of abusing and losing Marilyn in the process of making her a representation. The loss of her personhood is the eradication of her own thoughts about all of this to begin with, but that is an eradication that began when she became a test subject, that is, when she entered the consumer stream. The equally haunting absence of her body suggests that she has been cut to pieces as a result, subjected to scientific manipulation as much as capitalization. We can say the same about the image of her as a representative of women, if we choose to see her in that more general way. In this interpretation, the artwork suggests that neuroscience imagery shows us the gender stereotypes/constructions that we already want or expect to see, which is to say that the work announces that neuroscience cooperates with existing social and psychological inscriptions, whether or not anyone at Berkeley or Max Planck ever wanted it that way.

There is, as "Marilyn's Brain" can demonstrate, a deadly serious side to postmodern play. The social ills and psychological disquiets that cannot be stomached spew out existential disturbance in a parody. It functions like a vomit gag during a comedy show. But the audience might get sick from watching the splatter. Jean-François Lyotard says something similar in his discussion of postmodernism:

> The postmodern would be that which, in the modern, puts forward the unpresentable in presentation itself; that which denies itself the solace of good forms, the consensus of a taste which would make it possible to share collectively the nostalgia for the unattainable; that which searches for new presentations, not in order to enjoy them but in order to impart a stronger sense of the unpresentable.[37]

The brain art that presents the unpresentable of neuroscience—the unattainable of the brain or the problems and losses inherent to brain science—becomes a Critical Neuroart. Within this decidedly postmodern way of proceeding, neuroscience is seen as offering no solace in form. "Marilyn's Brain" situates neuroscience itself as the performance of postmodernism: nothing new, recycling social stereotypes, another hybridization, another fragmentation of the Self.

Yet "Marilyn's Brain" sits today uncomfortably on an artistic intersection of sensibilities: boredom with old scientific realisms mixed with the triviality attached to post-modern performances. Brain art, accordingly, can really only be successful if finding an unnerving combination. The conformities of the institution must get smashed with the comedies and parodies of the social real. In this boom stylization where the data-subject is a brain scan, the colorful brain artwork might raise from the dead a "refusal to provide explanations" after the death of "grand narratives," even while selling us a grand narrative that gives us a feeling of new life.[38] Some brain artworks more than others, of course, will cling to the idea that "society is a spectacle," but only as long as the spectacle has beneath it some actual neuroscience about who we really are; only then will it, cynically, catch the contemporary eye and do the work that makes us think.[39]

In spanning Modernism and postmodernism like a comic character having each leg balanced on two logs that slowly drift apart, playful brain art generates both pleasure and confusion. It offers something beautiful yet bureaucratic. It happily contradicts the idea that art today produces only cheap images that reflect a "culture of the depthless image"[40] even while making clear that a brain map is composed through a process where what is easily enacted at the level of experimentation is what is acceptable, socially expedient, and inexpensive. (Here, the description of the "cheapened" image takes on a double meaning.) Making participants read limericks in a brain scanner, identify breeds of dogs, think about rap music, watch a football game, listen to Kenny G, or read passages of the Bible are experimental tasks making the real of neuronal firing confusingly social and playfully light-hearted. The neuro-critical brain art in the museum, when foregrounding cultural engagement, shows that the image is not "depthless" even if it loses sight of the brain *as it is*, imaging nothing separate from social enactments even while declaring separation.

Standing in the museum staring at a big glowing tangle of neurons flashing like a Christmas tree amuses us. I am reminded of a scene out of David Grossman's novel, *A Horse Walks Into a Bar*. The audience gives an uncomfortable laugh as the comedian gyrates crazily on stage pretending that he is a teenager resting his head against the window of an army truck and his brain is getting rattled; the slightly crazy army driver tears across the desert, over pot holes, across gravel, assigned the duty to take this kid to a funeral, all while his head jiggles and crashes around the truck.[41] Grossman offers this description of a comedian acting out this scene:

> His features blend together, expressions cross over each other in flight like cards in a deck being shuffled. His limbs flutter and dance as he jerks around the stage, tossed from one edge to the other … "*Drrr!* Can't think of anything, don't feel anything, every thought gets crushed into a thousand pieces, I'm thought-paste! *Drrr!* … all of a sudden in the truck, my brain shows me a movie, don't ask me why, brains will be brains, don't expect them to be logical."[42]

Playful brain art puts the audience in the truck and locks the door. Who knows what movie we will see? How many movies are we made of? Maybe something starring Marilyn Monroe. Maybe *Dr. Strangelove*.[43] Maybe an old cult classic: *The Brain that Wouldn't Die*.[44] Neuro-critical brain art can be likened to the comedian's performance. And like when witnessing that performance, we eventually have the uncomfortable realization that the comedian on the stage is the same teenager that he performs; the performance then is about the loss of his own mother, and we want to laugh and cry at the same time.

Scan 14: The Brain of the neurosciences repeats correlationism

Conceptually, the turn to neurons to explain our intimate mental lives is akin to an acrobatic reversal. Simply put: when thinking our Self as The Subject, we make the brain The Object. But if we think of ourselves as a brain, then the brain suddenly takes the place of The Subject. In this case, the object that composes the Subject has to become the material stuff of brains, or neurons. Here is where the formulation grows more difficult—because we have a little bit of trouble thinking about ourselves as neurons. That can seem cold or just plain reductionist. But the trouble goes yet further.

As Sprevak notes in his "neural insufficiency thesis," we live not only with things but through them, and this means that we think, quite literally, through pens and computers and caffeinated coffees. Accordingly, complex psychological problems, although dependent on what's in the skull, continually take shape from everything outside in a process of engagement. As Sprevak puts it,

> We have no reason to think that the mechanisms involved in psychiatric disorders are entirely neural, and that fMRI, or even a perfect science-fiction brain-scanner, would be capable of uncovering them. Psychiatric disorders, like numerous other cognitive processes, are liable to cross the brain-world boundary in such a promiscuous way as to be resistant to neural reduction.[45]

Accordingly, we must turn the explanatory boat back around. In brief, to make the brain the Subject when neurons cannot really be the Object for that Subject, we have two choices. We can either (1) look to the whole world of interaction as the Object or (2) reverse course and make the Object of the brain (as Subject) an abstraction, which we will call in this book The Brain.

Because individual neurons (as Object) cannot fully account for the complexity of the experiences of The Subject, we need to refer to the whole brain again. And then we have a circle. Consequently, this time, the whole of the brain must be abstracted—The Brain—which is usually some ideal neuronal collective or an ideal computation that composes the conscious Subject. Forcing the webby insides to rely on an abstracted, idealized form called The Brain means that one can lean on the other for explanatory power. The little brain of neurons can become a Subject if pointing to its idealized formation, The Brain, as composing its life. Likewise, The Brain that is the abstraction can point back and say that it is a collection of neurons, a network of flesh. When one points to the other, it is all there. The high flipping acrobat sticks the landing and raises arms triumphant. We want to applaud.

But there is still a problem. The Subject and the Object—the Self and the brain, or the brain-as-Self and neurons, or neurons-as-Self and The Brain—are poorly acquainted. The Self and the brain only have so much in common; they really prefer to hang out at different parties. The brain has a party that is mostly flashing lights. The Self prefers long conversations and classical music. Some people say that there is a back-door entrance from one party to the next (like the beer kegs are actually connected or they smoke the same pot or some such thing), but neither side seems willing to admit the truth.

Likewise, the brain knows little of The Brain. Neurons go out to their party but only ever dance to pop music and suck down gin cocktails. The Brain searches for its tuxedo amongst all its little neurons, but they are always elusive, too busy or too snobby to explain the whole picture of themselves to some snob like The Brain, who hangs with the celebrities and lives in Hollywood.

Put less figuratively: we can imagine that the materiality of the Subject and that of the Object, as philosopher Quentin Meillassoux argues, remains at points independent, always something more or else, always changing and exceeding a boundary imposed by an outside. The Object, in his view, sits ever independent of one's position as Subject (i.e. they don't talk very well or go to the same parties). As he states, "it is possible to consider the realm of subjectivity and objectivity separate from one another."[46] The Subject cannot be encapsulated by any Object. This would include the brain, of course. If the world exceeds correlation, then our brains may be other than what we are—so much more! In other words, if Meillassoux is correct, then we are aliens to ourselves, inexplicable Beings. Are we not?

How, in this case, is the world to be known? For Meillassoux, the answer is speculative imaginaries, by which he means building interventions, not limiting ourselves to sense perception but dipping into abstraction and experimentation with materials to manifest (generate) something not-yet but still conceivable, at least conceivable after its appearance.[47] The whole idea of speculative imaginaries is based on the notion that "there are no necessary relations between observable instants," and, as such, the approach dedicates to novelty, or what Meillassoux calls "virtuality" in a probabilistic sense of the universe's mass of actually possible emergences.[48] Thus, Meillassoux crafts an ontology wherein the world's materiality is aligned with the desire for objects to become always new, for life-worlds to evolve around us, maybe for our wishes to come true. Or our worst fears. Indeed, in being probabilistic, all hopes are attenuated.

Framed in terms familiar to art, Meillassoux's ontology expresses a dedication to invention. What more could any wild dare-devil artist want but this ontology? On an individualized and discrete scale, any Object might suddenly exist in whole other worlds, or not be there anymore at all. Objects might suddenly be more relatable and then not. This maniacal philosophy can be likened to an artist giving the materials of an artwork over to chance. Thus, the ontology can remain void of the kind of human-centric art-novelty that valorizes the artist because something was said to be shaped with genius, yet the art can still make "a significant contribution to a shared memory."[49] The prioritization of novelty inherent here is also a prioritization of unpredictable Being for individuals and communities.

We can read this from how we feel about it.[50] That's one way to start anyhow. In prioritizing unpredictability, the ontology that Meillassoux creates is itself probabilistic yet forwarded as a description. If truly probabilistic, then we might suggest that Meillassoux is lucky to happen upon such a view and be so situated. That is a first reaction, which is my own gut reaction. But after grasping his Speculative Materialism analytically, we might discover, yet again, a pitching feeling of sadness; sensing what it means for us to be human from inside a commitment to sudden articulations, resonances, capacities, radical divergences, instantaneous emergences bring about whole sets of confusing feelings. The uproar of the unknown stirs the soup of the guts; molten lava is the solid earth's ironic core.

But breaking the Object from the knowing Subject is precisely to stop feeling so sad for being so out of control of human futures in this strange planet just as much as to stop feeling so special for being so brilliant about controlling it. In fact, Meillassoux tries necessarily to step past the emotional and psychological trauma of correlationism proposed by thinkers like Sartre and Heidegger; he seems to want to think beyond a history of existentialism wherein the subject reaches out to know and then finds an inescapable hiddenness or emptiness at the core of all Being. But does he succeed?

If his ideas succeed by presenting an excitement to living or if they offer reasons to keep on living—to invent, to be invested in the potentiality of Things—and to adopt a realism without correlationism, then I think that it also fails pretty miserably at resolving any confusion about human life; the same can be said if this materialism is seen as an attempt to resolve despair. Thinking through the implications of Meillassoux's proposal does little to banish the feeling that we confront conditions of treachery. We might even say that Speculative Materialism hopes too much for materiality and, thereby, worsens the existential state. When materiality is so wild and probabilistic, so are any felt successes. But even more damning, the mathematician knows that there is no end to infinity.

Neuroscience, however, tenders a different case.

Brain findings offer compelling accounts of brain processes, and nobody doubts that the brain is central, or at least necessary, to most mental phenomena. Neuroscience convincingly relates observations of active brain systems to human experience, and in this, neuroscience feels so much more real and dependable. Structures of the brain and the attendant formations of mind are not left to chance. The Subject and the Object are not living wild parties divorced from each other. They are correlated, replicated, and made durable in an epistemic apparatus. They are nerds hanging out together in a computer lab.

In fact, the human constitution promises not to exceed our capacity to understand it once the underlying physics and genetics of the brain have been fully detailed. Nature's Laws provide psychological stability. Biomedical and scientific tools provide means of intervention. Scientific realism is tailored to compete for hearts as it elucidates the mechanisms of minds.

If we are honest, however, then we quickly remember that neuroscience, too, suffers from the same problem that Meillassoux identifies in the first and then tries to solve. That is, the brain as Object may well exceed the boundaries of the participant or the neuroscientist as Subject. The brain shoves the philosopher out onto the edge of the ship's plank, over the water of uncertainty, threatening to push her over the edge. She turns back to see who is doing the shoving: it's herself!

Neuroscience is trapped in correlationism.

It is an intriguing case because the Self can rhetorically occupy both positions. As Subject, we can come to know what is real about the brain as Object through the workings of the brain. So the brain becomes also the realist Object of the Subject—but is made real by the Object. In other words, the brain is that which we must study to understand how we come to know what is real; however, the brain is also the Thing (we cannot but now believe) that makes all that we see *as real*. The Subject looks at itself even while not itself really knowing whether the thing positioned as itself lies.

Inter-reliance is the difficulty of correlationism in neuroscience. But neuroscientific inquiry needs Subject–Object correlations for claim-making. Here, we can remember the radical proposal that Meillassoux makes. If Objects are, as Meillassoux suggests, active in realms far beyond any correlation and able to have viable, functional aspects without a sensed correlation, then science will inevitably look sometimes like that dog barking up the wrong tree. What neuroscience identifies may have little or no relation to the correlated entity that it claims to see and discuss. Likewise, we, as test subjects, might do some barking ourselves, but we might even be running in the wrong forest. What we *are*, in other words, might not be accessible, not even to us; the interpersonal "I" could have no correlation, we can at least imagine, to those aspects able to be studied by neuroscientists.

Neuroscientists are not ignorant of this philosophical problem. In fact, they do something about it by doubling down on inter-reliance. By pursuing one (or both) of the two solutions that Harman says characterizes most philosophies—overmining and/or undermining—neuroscientists effectively navigate the criticism that correlations prove insufficient.[51]

How does it work? Neuroscience "overmines" by reducing down the brain to neurons, synapses, and molecular components—all those things that are more tangible than the whole organ. The brain then is then forwarded as a series of material relations; thus, this brain-organ is able to be seen and tested. However, the mind in the brain can also remain hidden within the massive complexities of relations needed to sustain the viability of overmining. As a result, conversely, neuroscience can choose to "undermine" by reducing up the neuronal brain to The Brain of overarching abstraction, especially when relations do not seem to explain what is happening. The undermined The Brain is a meta-concept composed of the discipline's unifying (and usually computational) ideals, which serve as the more real/most real interpretation or conception. The Brain is an ideal that can then be tested computationally (and perhaps in other ways) and put back into correlation with sense perception.

In practice, neuroscience can reduce down to explain the abstraction (it's neurons), or it can abstract up to give reason to the reduction (it's an ideal material form/ation, a particular kind of emergent computational form, a paradigmatic consciousness). Turning in circles ensures that the senses and good sense can still tell us a reasonable story. But none of this, we can now see, resolves the original epistemological problem (which is ontological at its root, of course) nor the existential bothering.

When neuroscientific knowledge-making requires a series of explanatory appeals turning in a loop, then the space for Meillassoux's ontology (his "break" in the correlationism thesis) opens up. Exploration of potentiality then works toward invention or crafting for other kinds of answers. The problem caused by The Brain that cannot explain a brain's neuronal firings without connections to environmental relations, just like the brain that cannot be related to experience at all without a disciplined ideal working behind the scenes can be artfully handled through asserting some non-co-dependence, i.e. the end of correlating a Subject and Object. But does it pay off for us experientially?

The admirable trait of Meillassoux's Speculative Materialism is the total abandonment of sense perception as final word and then the invention of new imaginaries that might re-assert realism without any guarantee. In kind, the admirable trait of rejecting this wondrous materialism is situating it as an attempt to maneuver cunningly around the body, i.e. not allowing clean and easy passage free from tears and the tearing out of hair for any claim that the world can be suddenly other than what we see, what we know, and what we feel when all we have is a body. The seemingly most viable alternative, of course, is correlationism. But why don't we play around a little?

Performative play with neuroscience

The human brain can be a raucous place. Who knows what goes on in there. We can't keep track of it ourselves. In the span of a minute, we can have memories of catching a bluegill at grandpa's farm. Insecurities about sweaty armpits from high school haunting us at the office. A flash in the corner of the eye, not quite registered. A burst of insight in the middle of a business presentation. Then an irritating feeling that we should be somewhere else—Oh yeah! We had an appointment! Shit!

Choreographer Jody Oberfelder's "The Brain Piece" makes this unending flow of continuous dealings and divergences a property of the mind tout court, even though it may be a case of mind meets Modernism, urbanism, technologies of distraction, labor demands, reorganizations of family responsibilities, and more. "'Sometimes the mind is really chaotic … So I wanted to create an environment where you're completely distracted all the time.'"[52] Although this kind of environment sounds suspiciously like the average American family living room where teenagers and parents bounce between mobile phone apps, video games, emails, stove tops, homework, phone calls, Netflix shows and more, Oberfelder seems uninterested in blaming anyone but the brain for this problem of mind.

As a dance performance, "The Brain Piece" captures the kind of jarring, often unexplainable, character of everyday life where we find ourselves thinking about a grocery store list and then suddenly crying at a puppy being rescued from a raging river in the newest viral video. Dancers fly across stage. They gyrate like electric neurons. Then everyone stands and joins in a game of tug of war between the so-called Left and Right brain. Suddenly, the lights explode, and there are dancers projected across the walls and floor. Their arms are raised above their heads; they spin, looking like a bowl of SpagettiOs. In the next moment, the dancers, about twenty-seven, join arms with the audience. They network to form a living brain. Each person pushes and pulls, reacting to the touch of another, and everyone gets the feeling that the brain is a body or the body is the brain and everything is interconnected.[53]

In viewing "The Brain Piece," we learn that thinking about our own bodies shifts when putting the body to work, testing it out, taming it, letting it swing free. The sheer experience of movement, as dancers tend to know, inherently asks questions about the independence of the mind. Suddenly, the fluidity of a bow, the roll of the torso, or the stretch of the legs, relaxes us. Mental reservations that used to hold us back start to dissipate. Soon we are dancing. Whatever worry suffocated us only a moment ago is given over to a new musical freedom vibrating in every limb. But somehow, bodies still get the short-end of the stick even if we refer to them inherently when talk about how we feel; strange how the brain can steal the credit.

It is not easy to scream profanities at a tablecloth. Perhaps we view our arms and feet in the same way. The brain, we believe, deserves the force of the derision. This dynamic, where the brain always gets or takes the priority, is embedded inside of Oberfelder's own description of the dance performance. She voices the contemporary prioritization of the brain even as she showcases bodies in living relations. For example, she describes how a previous work exploring the rhythms of the heart (called "4Chambers") led to the realization "that everything in your heart, everything you feel — it's all the brain."[54] When describing "The Brain Piece," she cannot ignore the co-constituting role of the body in thought production, but it has its place. She states, "I just think dancers, and in general physical people who integrate what they're thinking through the body, are the most brilliant people."[55] In this quotation, the body follows after-the-fact and is integrated into a trajectory of thought. That even a professional dancer, hyper-attuned to bodies, has difficulty escaping the linguistic hierarchy of the brain says something pointed about the contemporary order.

Oberfelder's lack of clarity on the brain–body relation, and maybe her hesitancies to say that the brain is thinking *from the body*, are mirrored in the work's semi-structured performance. At key moments, the audience is invited to jump up and participate (a choice that is left ultimately to the viewer's own discretion), while other times, the audience must sit and watch. Oberfelder's dance composition therein reflects the tension between the structured unities of the neuroscientific stage and the open-air wildness of bodies that eat and breathe bacterial creatures and viruses. In the dance, this tension is experienced as a well-ordered progression followed by totally unstructured performative reactions. Thus, despite giving some privilege to the brain—strengthened by Oberfelder's appeal to her consultations with a neuroscientist to lend credibility to the dance movements—the communicative insights of the work leap well beyond neural discourses.[56] That is, the performativity of the work—the sudden swell of bodies, the synchronization with rhythm—initiates skepticism about the confinements of the brain. The dance quickens speculation about the brain's unities with the body and environment. In fact, watching all of these bodies shift variably to the flash of colors, react to the crescendo of the music, fall to the turn of the digital screens, beat against the hardness of the floor, and swoon with the hot breath of another right up on the neck removes the brain totally from sight. What is seen, or experienced, rather, is a suturing of the mind–body. What is felt inside of the theater is a nothing-brain that cannot exist at all as anything without embodiment of the ambient space.

The brain can do nothing of its own. The totality of "the we" becomes visibly evident as a mesh of hands and feet and eyeballs and lights and screens and the theater. The theater, if anything, should be called the brain. The brain, if anything, should be the all-encompassing moment: the pop of a drum, the clap of a heel, and the rush of blood to a face. The absence of a brain is what we encounter in "The Brain Piece." To talk about this work as "The Brain Piece" feels like a giant misnomer.

As a speculative dance, the unexpected moment open to the audience's participation—personalities, sizes, strengths, etc.—is where the dance claps out the best rhythm. Through the unexpected moves of the audience members, the performance releases itself from the rigidities of choreography. The individual dancers who ostensibly represent neurons must let go of any hope for precision; they must dare not to pretend to be in control of everyone's experience, and they cannot act as a totalizing or territorializing force. The dance requests something more. It opens itself up to intimacy with Others and presents an ethics of communion where nothing, really, is paved in advance.

We might balk here, briefly, and argue that these moments of freedom and non-Modern performativity are, themselves, staged amid a broader choreography. However, they nevertheless communicate an important point. They jolt the mind to turn toward the instantaneousness of a brain enmeshed with the wider world. They highlight the ceaselessly generative and display a body out of itself all the time. They compel us to reverse a geography where neurons are acting and not being acted upon. As such, the brain is not at all what is staged here.

Scan 15: brain art is madness

Even if surveying the amazing innovations of the entire Italian Renaissance, Leonardo da Vinci probably never could have imagined such ripe conditions for bringing art and science together as exist today. What better place to direct the power of artistic invention—a shaping of materials, as my artist friend Jason Kofke often says[57]—than a brain science setting out to articulate why we love to drink whiskey,[58] why we vote liberal or conservative,[59] why we fall in love,[60] and why we believe in God?[61]

Visualizing the intermingling of local narratives and the brain sciences, or dressing neurons seductively, or calling audiences to look "at" and not "through" a brain scan,[62] presents a series of compelling artistic tasks as people struggle with states of matter and the extent of material determinism (or lack of) over their lives. Brain art, in other words, can reflect what seems frail or impressive about bodies, what seems old or radical in the sciences, what seems right or wrong about limiting our talk about Alzheimer's disease to a problem with tangled up nerve cells. And brain art can help to achieve the task of complicating any of the binaries simmering there (frail/strong, good/bad, right/wrong). I like to believe that brain art—specifically, artworks depicting brain discourses as much as brain systems—holds the potential to compose alternative ways of imagining ourselves simply by bringing attention to how we already do.

If art is at its core "madness," as Henry James suggests,[63] then reimagining and playing with the wild task handed over to the neurosciences, as something both sane and insane, makes good sense for brain art. Indeed, any attempt to chart conscious–unconscious, affective–cognitive unities appears immediately sane and yet so difficult as to be insane, especially once scoped down to investigations of a single organ. The artist is liable to prove at times astonishingly cogent but at other times distressingly incoherent.

The lines between sanity and insanity, daydreams and drugs, miracles and moods, eccentricities and erratic behaviors, eroticisms and reason have long been crisscrossed by artists. The lack of stable readings (of artists and of their artworks) adds to their historical endowment as geniuses. As Neihart notes, madness is expected of creatives both in the form of their art—because it exceeds the imagination in breaking visual norms—but also because artists are often willing to engage the taboo and give life to animalistic desires.

> Since the time of the Greek philosophers, those who wrote about the
> creative process emphasized that creativity involves a regression to more
> primitive mental processes, that to be creative requires a willingness to
> cross and re-cross the lines between rational and irrational thought.[64]

Performing the unreasonable position that the brain controls the ultimate
real for us yet can provide for us some certainty about the shape or effect of
materiality makes brain art—inhabiting that ambiguous middle space while
playing both sides simultaneously—madness.

Brain art, in its touching and tantalizing conglomerations, its attractions
and sideshows, its techno-illuminations and neuronal obsessions, its maps and
computer metaphors, its luster and lingo, necessarily reflects a foundational
human desire to know Things through all kinds of untidy, untrustworthy,
lovely, and turbulent human means. Wood, glass, metal, magnets, LED lights,
flame throwers, beakers, breakers, threads, and circuit boards. All piled up to
say Hello World. In this messy conglomeration, brain art reflects an ignorance
about how to know Things, how much we can detect, whether we can
reason, and why Things might mutate or stay the same. Art taking the brain
as its object chooses an adventurous and sensual un/known, a blend of right
here and way out there. Brain art is a kind of beautiful madness and may well
inject some sanity into the maddening endeavor of finding a right means to
know about ourselves.

Performativity need not be dozens of twisting bodies on a downtown stage. Creative, regenerative, open-ended engagements also foster new ways of approaching
what is otherwise too stupidly solidified. Crafting experimental conditions where
the unexpected explodes mindless simplifications and rote answers makes a mockery
of genre while bringing some clarity to the elements needed to constitute it as such.
Working with surprise amid what seems religiously regimented, and thus terribly
unsurprising, is a mode of critical intervention. In fact, to play amongst material
relationships and to encourage social and cultural analysis by remaking stock phrases,
by slicing up stodgy scientific models, or by rehashing the taken for granted arguments point by point except for one crucial detail wildly refashioned, steps out to
upset ideological formations. Performativity gears up for this adventure by rustling
up combinatory possibility.

I try to get my hands dirty with paint, glue, coffee, connectors, code, what-have-
you to practice what I preach. In thinking through what a performative and critical
neuroscience can do, I constructed what I call the "Neuro News Generator."[65]
This is a re-combinatory new media artwork made with the Processing program-
ming language. The piece presents a cheeky parody of popular neuroscience news,
aiming to intensify "awareness of the social implications of research and its uses."[66]
On screen, viewers see a series of popular neuroscience headlines with a word or
two missing. The "Neuro News Generator" is programmed to randomly generate

new "hot topic" words to complete the formulaic headlines, foregrounding recurring topics, demonstrating how stock phrases can uncritically stage neuro-science as the final word on the realness of some habit or belief. For example, the sentence, "Neuroscience shows _____ really does make you happy" is auto-matically completed with words like "television," "chocolate," "exercising," "good sex," "golfing," and others. The sentence changes every three seconds. In making this work, I perform a neuroscience confirming existing folk practices and demonstrate the way that findings can be so easily adopted as consumeristic pleasure or trifling Self-authorization and get wrapped up in pre-existing expect-ations about the so-called good life.[67]

Artist and critical technologist Daniel Howe offers another example to consider. He works on a wider scale, integrating new media functionality and probability as a form of creative institutional and cultural interrogation. In a recent work called "Advertising Positions," Howe teams up with colleagues to create "software robots trained to search the web, according to specific user profiles."[68] Over a period of weeks, these online robots, like little consumer spies out on a mission, visit numerous websites fitting to their profiles to collect the advertisements. Each advertisement is then applied "to a single polygon on the mesh skin of the virtual avatar it inhabits. The training period concludes when the robot's 3-dimensional figure is complete."[69] The result is a body whose skin is made of Mercedes, Garnier, Whole Foods, and Prada.

Because advertising reliant on the collection of user profile information and behavioral data is sold, shuffled, layered, and shared, the work displays a body that is both private and public. We encounter a hybrid body ever shifting with media corporations and consumer apparatuses that target it for persuasion. In looking at the giant logo stretched across the chest, we consider how we are compelled by consumeristic automation to buy more, stop here, go there as we interact with Facebook, CNN, YouTube, and pretty much every mobile app. Our visits are aggregated. Our maps are tracked. Our email messages mined. Our shared photos analyzed for location and time stamp. Our facial recognition software always on, learning the contours of our cheek bones. Algorithms touch our bodies like blind lovers, groping to come into some understanding of what we look like, what we want, what we desire. As Roberto Simanowski notes, they aim to woo us, but they prove to be selfish lovers only out for themselves.[70]

Slathering a body with advertisements exteriorizes the needed attention to the lack of privacy tucked into complex and shadowy algorithms driving new media environments. In Howe's work, the body is naked but clothed in an excess. The body screams. The advertisements fold across the neck, over the face, and then roll into the mouth. They fill up the inside, which somehow remains hollow and black. Whose body is this? Is it being remade from the outside in or from the inside out? Then, a shift. Another ad from Starbucks, another line of code from Airbnb, another offer from Cathay Pacific. Like the hidden computer systems that do the fast aggregation and content selection, the transactions that build the body being displayed on screen "occur at light speed while the page is loading."[71] We realize how inseparable we actually are from a vast technological apparatus stretching from Singapore to Silicon Valley to a quiet outcrop at an NSA warehouse in Hawaii.

A work like Howe's captures the salience of performativity in thinking about bodies. And the work could easily be reprogrammed to entertain dimensions of the contemporary neurosciences. In a spirit of adventure, we can consider a few possibilities.

Images applied over popular neuroscience news headlines could be collected the same as advertisements. Similar to Howe's personality profiles, they could be divided by topic, showing the images featured alongside the so-called "adolescent brain," "Autistic brain," or "aging brain." Images could be further divided based on whether they originate from the neuroscience lab, from a news publishing site, or from a stock images website. They might then be displayed in any number of ways, perhaps, over the three-dimensional frame of an adolescent body or an aging body. What would we see? A human body plastered in brain imagery? Or social symbolics both disturbing and alluring?

We can push the idea further. Imagine building algorithms that matched images of known psychopaths with brain scans labelled as psychopathic brains. Or, we might generate an algorithm that creatively recombines facial profiles with brain scans and then asks well-known neuroscientists to determine what they see. The results could be compared against the general public's determinations. We might throw in a picture of your sweet grandmother. Or maybe Mr. Rogers. Or that creepy basketball coach that you feel the need to keep an eye on after practice. In terms of knowledge production, what would be scientifically valid from such works would be of significantly less importance than whatever the audience is compelled to do, to rethink, to question. That would be the point. Performative works such as these ask what prejudices emerge, what connections are too common, what alternative trials might be composed, what possibilities can now, after the interrogation, be invented. And they do so through the unexpected combination, the inculcation of the strange in the familiar.

Releasing intended relations

To my mind, a totally new brain art is material un-formation. A pursuit of open-endedness in neuroscientific exploration means making brain art that reveres a "white silence … pregnant with possibilities."[72] This altered vision of the way far out, seeking the blurry ends of the earth, a material exchange in an expanse, blends the crisp visualizing capacities of the neurosciences with experimental art practices that romp amid agency and matter to intensify how actively Things entwine to suddenly Become what was not scanned and could never have been imagined.

If present neuroscience–art projects tend to re-stage the old social categories and offer historical pictures as images of the future, then they probably adhere to the scientific impulse to locate and define, ignoring the insurrection of the genera-tive. If they shake hands nicely with Descartes, then they should try slapping him for once, or maybe stick a big middle finger in his face. If they play gracefully in the orchestra of the ordered Enlightenment, then they should now try jamming with some wild experimental jazz.

I propose a Critical NeuroArt as one possible avenue for expanding imaginations of the brain beyond The Brain of Power and Might. In Critical NeuroArt, neuro-sensuality and Affective Realism are not themes emerging from humanistic engagement. If there at all, they are activating means to make audiences feel with and through domineering neurobiological narratives set as explanations and guides for human existence. But more than this, an intense focus on iterative combinations and unnerving performances can, I believe, help to shake out alternatives for the brain and release the intended relations. In a performative embrace, neuroscience is retooled to act out what it might come to mean once understood through a lens of power and operating only with/in the body. Critical NeuroArt pounds the chest and slaps the thigh. It radiates what a body with a neuroscientific brain can and cannot do—and then it leaps out into a new space to explore what a body without The Brain can do.

Notes and references

1 For discussion, see Michel Foucault, *This is Not a Pipe: With Illustrations and Letters by Rene Magritte*, translated by James Darkness (Berkeley: University of California Press, 1983). Available at: https://archive.org/stream/FoucaultMichelThisIsNotAPipe1983/Foucault_Michel_This_Is_Not_a_Pipe_1983_djvu.txt.

2 Andrew Pickering, *The Cybernetic Brain: Sketches of Another Future* (Chicago: University of Chicago Press, 2010), 17.

3 Gilles Deleuze, *The Fold: Leibniz and the Baroque*, translated by Tom Conley (Minneapolis: University of Minnesota Press, 1993), 4.

4 Ibid., 19.

5 Ibid., xiv.

6 Ibid., 123.

7 Ibid.

8 Pickering, *The Cybernetic*, 8.

9 Jody Eberfelder, *The Brain Piece*, *Jody Eberfelder Projects*, 2016. Available at: www.jodyoberfelder.com/thebrainpiece/.

10 See Diane Coole and Samantha Frost, *New Materialism: Ontology, Agency, and Politics* (Durham, NC: Duke University Press, 2010), 1–6.

11 Karen Barad, *Meeting the Universe Halfway: Quantum Physics and the Entanglement of Matter and Meaning* (Durham, NC: Duke University Press, 2007), 141.

12 Thomas Rickert, *Ambient Rhetorics: The Attunements of Rhetorical Being* (Pittsburgh: University of Pittsburgh Press, 2013), xii.

13 For more on the overlap between ontology and epistemology, see: Annemarie Mol, *The Body Multiple: Ontology in Medical Practice* (Durham, NC: Duke University Press, 2003).

14 For more on this topic, see Gruber, "There is no brain: rethinking neuroscience through a nomadic ontology," *Body & Society* 25, no. 2 (2019): 56–87.

15 Ibid.

16 Kimihiro Nakamura, Wen-Jui Kuo, Felipe Pegado, Laurent Cohen, Ovid J. Tzeng, and Stanislas Dehaene, "Universal brain systems for recognizing word shapes and handwriting gestures during reading," *Proceedings of the National Academy of Sciences of the United States of America* 109, no. 50 (2012): 20762–20767.

17 P. Read Montague and Gregory S. Berns, "Neural economics and the biological sub-strates of valuation," *Neuron* 36, no. 2 (2002): 265–284.

18 Christoph Redies, "Combining universal beauty and cultural context in a unifying model of visual aesthetic experience," *Frontiers in Human Neuroscience* 9, April issue (2015).

19 Nikolas Rose, "Human sciences in a biological age," *Theory, Culture & Society* 30, no. 1 (2013): 21.

20 Shutosh Jogelekar, "Why the search for a unified theory might turn out to be a pipe dream," *Scientific American, Blog*, May 3, 2013, paras 1–2.

21 Clement Greenberg, "Review of the Exhibition Collage," *The Nation*, November issue (1948): 612–614.

22 David Joselit, "Notes on surface: toward a genealogy of flatness," *Art History* 23, no. 1 (2000): 20.

23 "'Marilyn Diptych,' Andy Warhol, 1962," The Andy Warhol Society for the Visual Arts, Inc. Available at www.tate.org.uk/art/artworks/warhol-marilyn-diptych-t03093.

24 This paragraph draws from my article in *Leonardo*. See David R. Gruber, "A Critical NeuroArt for Critical Neuroscience," *Leonardo* 29, no. 2 (2019): 1–7.

25 For a discussion of sexism in science, see: Vandana Shiva, *Women, Ecology and Survival in India* (London: Zed Books, 1988); Muriel Lederman and Ingrid Bartsch, *The Gender and Science Reader* (London: Routledge, 2001).

26 Ernst van Alphen, *Art in Mind: How Contemporary Images Shape Thought* (Chicago: University of Chicago Press, 2005), xiv.

27 For discussion of Warhol's sensationalism, see: Patricia Lee, *Sturtevant: Warhol Marilyn* (London: Afterall Books, 2016), 40–42.

28 Kelly Oliver, "The severed head: capital visions (Review) by Julia Kristeva," *The Comparatist* 37 (May): 324.

29 Ibid.

30 "The most important art in Modern art," *The Art Story* (2018): para. 7. Available at: www.theartstory.org/definition-modern-art-artworks.htm#pnt_3.

31 "Art: dripping with genius," *Time* 152, no. 19 (November 9, 1998): 84.

32 Holland Cotter, "Jackson Pollock at the Guggenheim: works of swirls and pixie dust," *New York Times*, May 26, 2006.

33 See Jeff Koons, www.jeffkoons.com/.

34 Oliver, "The severed," 324.

35 Ibid.

36 Ibid.

37 Jean-François Lyotard, *The Postmodern Condition: A Report on Knowledge* (Minneapolis: University of Minnesota, 1984), 81.

38 A similar dynamic is discussed by Simanowski in his book on new media art. See Roberto Simanowski, *Digital Art and Meaning* (Minneapolis: University of Minnesota Press, 2011), 181.

39 Fredric Jameson, "Foreword," in *The Postmodern Condition: A Report on Knowledge*, by Jean-François Lyotard (Minneapolis: University of Minnesota Press, 1998), vii–xxii.

40 Andrew Darley, *Visual Digital Culture: Surface Play and Spectacle in New Media Genres* (London: Routledge, 2000), 192.

41 David Grossman, *A Horse Walks Into a Bar* (London: Penguin, 2017), 127.

42 Ibid., 125–127, italics in original.

43 Dr. Strangelove, or: how I learned to stop worrying and love the bomb, 1964, Stanley Kubrick, IMDb. Available at: www.imdb.com/title/tt0057012/.

44 The Brain that Wouldn't Die, Joseph Green, 1962, IMDb. Available at: www.imdb.com/title/tt0052646/.

45 Mark Sprevak, "Neural sufficiency, reductionism, and cognitive neuropsychiatry, philosophy," *Psychiatry and Psychology* 18, no. 4 (2011): 339–340.

46 Quentin Meillassoux, *After Finitude: An Essay on the Necessity of Contingency* (New York: Continuum, 2010), 5.

47 Ibid.

48 Levi Bryant, Nick Srnicek, and Graham Harman, "Toward a speculative philosophy," in *The Speculative Turn: Continental Materialism and Realism* (Melbourne: RePress, 2011), 13.

49 Julian Stubbe, *Articulating Novelty in Science and Art: The Comparative Technography of a Robotic Hand and a Media Art Installation* (Berlin: Springer, 2017), 214.

50 I say "we" here out of convenience and what I take to be common feelings; however, I do not presume everyone feels this way.

51 Graham Harman, *The Quadruple Object* (Alresfo: Zero Books, 2011), 6–11.

52 Siobhan Burke, "Does the body think? Do your neurons dance?" *New York Times*, June 27, 2017, para. 3. Available at: www.nytimes.com/2017/06/27/arts/dance/does-the-body-think-do-your-neurons-dance-jody-oberfelder-new-york-live-arts.html.

53 See Eberfelder, *The Brain Piece*.

54 Burke, "Does," para. 9.

55 Ibid., para. 6.

56 "Jody Oberfelder Projects: the brain piece," *New York Live Arts*. Available at: https://newyorklivearts.org/event/brain-piece/.

57 See Jason Kofke, jasonkofke.com.

58 Monique Beech, "This is your brain on whiskey," *MedicalXpress*, April 9, 2015.

59 Katie McNally, "This is your brain on politics: the neuroscience that shapes our views," *UVAToday*, March 2, 2017. Available at: https://news.virginia.edu/content/your-brain-politics-neuroscience-shapes-our-views.

60 David DiSalvo, "What neuroscience tells us about being in love," *PsychologyToday*, February 12, 2014.

61 Jordan Rosenfeld, "This is your brain on God," *Good Health*, December 21, 2016.

62 The distinction between looking "at" and not "through" was first coined by Richard Lanham. See: Richard Lanham, *Electronic Word: Democracy, Technology, and the Arts* (Chicago: University of Chicago Press, 1993), 153.

63 See: Joyce Carol Oats, "The madness of art: Henry James' 'The Middle Years'," *New Literary History* 27, no. 2 (1996): 259–262.

64 Maureen Neihart, "Creativity, the arts, and madness," *Roeper Review* 21, no. 1 (1988): 47–48.

65 See David Gruber, "Toward a critical neuroart in critical neuroscience," June, 2018. Available at: http://www.neurohuman.com/criticalneuroart/.

66 Suparna Choudhury, Saskia K. Nagel, and Jan Slaby, "Critical neuroscience: linking neuroscience and society through critical practice," *BioSocieties* 4 (2009): 62.

67 See Eric Racine, Ofek Bar-Ilan, and Judy Illes, "fMRI in the public eye," *Nature Reviews Neuroscience* 6, no. 2 (2005): 159–164.

68 Daniel Howe, Qianxun Chen, and Chen Zong, "Advertising positions: data portraiture as aesthetic critique," *Leonardo* 51, no. 4 (2018): 413–418.

69 Ibid., 413.

70 Roberto Simanowski, *Data Love: The Seduction and Betrayal of Digital Technologies* (New York: Columbia University Press, 2016).

71 Howe et al., "Advertising," 414.

72 Wassily Kandinski, *Concerning the Spiritual in Art*, translated by M. T. H. Sadler (New York: Dover, 1977), 39.

CONCLUSION

For the love of the momentary brain

Brain art as another possibility

Going to an art gallery and finding brains covered in gold leaf or neon glow-in-the-dark paint can leave the visitor a little bamboozled, suspended in a kind of delightful and confusing shock; the viewer wonders what it means and what is being interrogated. The answer depends on so many references and material forces. At times, the brain looks to be a lively pop-art throwback, flashy as Andy Warhol spinning in a disco fantasy, and equally invested in Capitalist hoaxes. In other instances, images scrupulously copied from brain scans strike the viewer as an essentialist reproduction of Leonardo da Vinci's greatest hits—best-selling songs include "Look Inside and Love Yourself" and "What a Beautiful Thing We Are" and "See Here, This Explains Everything." Many museum visitors seem happy to leave with an exciting exhibition program detailing the visual wonders of technoscience. But this would be to ignore the surprise and horror of some brain art.

The brain can be made into a disturbing neo-psychological realism, more intense than the scribbly face dripping with paint. Brain art can strike the viewer as more real, yes even *more real*, than the hyper-realism of Chuck Close.[1] Then, suddenly, brain art can transform: a soft pillow of emotionality. Intimacy can be made with scans; subjective experience can be pictured, we accept, as we gaze into a mesmerizing jungle of neuronal complexity. Neuronal lines inscribe a new sense of access to our lovers. Some works of brain art promise sweeter understanding about why we hold hands or why our partners sleep better when they wear our clothes to bed. We might even find another means to express mourning. Neurological images enshrine memories, or better put, serve as shrines to our memories, like when we discover grandpa's old watch in the back of the closet. Through them, we are able to remember the long, tiring days in the hospital.

Then we step back, reflect. We wonder what it is to be that brain captured and slapped onto the museum wall, whether we are that one glowing in bright red, whether we are not. Do we feel the same? We turn and ask if this image is meant to suggest that we are neuronally on fire. Does it make a claim on us as abnormal? Are we "normal"? Then we pause. Wait, is this a joke? The brain image in the museum might be playing a kind of "gotcha" game, we realize. Then we comfort ourselves with the observation that other people make the same connections and fall into the same correlation traps.

The psychological journey through an art gallery featuring brains emphasizes a point: brain art is an inward twist and an outward fold. Engaging it means reaching out with one hand and turning inward at the same time with the other. Brain art traverses both sides. We hold the scan and see how we feel. It goads the dilemma of embodiment—not being sure about our bodies, what they are, or whether our scientific inscriptions get us further down life's path for understanding how we feel. Abiding with brain art, and crucially letting its multiple interpretations sink in and envelop us, as I have tried to do throughout this book, becomes a coming to grips with knowing and unknowing, performs an ontological flip-flop, asks an enduring question.

The brain may be forcefully in the center of the picture most of the time, but the presence has an absence. The background of an fMRI scan is usually pitch black. It is difficult not to focus on the brain, the object of illuminated evidence; the blackness behind it looms. That thick, terrible nothingness must be noticed; it does not care for its context. The absence, like outer space trapped inside of the skull, punctures the image, clutches our attention—there seems to be more back there than we could ever see. We tumble into the picture. That grave impenetrable space. We wonder how much we really know.

Returning to the trajectory of thought at the end of Chapter 4, I wonder if the entire canvas of neuroscience could ever be painted black. What about a "white on white" painting of the brain? Will we ever have a Kazimir Malevich or Robert Ryman of brain art? Can the brain lose its surface, discard the figure-ground relationship in the image? Can the brain image reveal the traces of the painter and be regarded as technically manufactured? Can the flattened and hopelessly decontextualized still be seen as a brain? Yes, yes, I think so. That is precisely what we view in images of a frontal cortex divorced from any environmental context. Those images extract the blood and guts, chemical signals, eyeballs, tongues, magnetisms, algorithms, florescent lights, and nice researchers whispering to patients stuck in fMRI tubes, "Now just breath and focus. This will be over in one minute." It is surprising that this compressed object does not suddenly appear to us as a parody of our own neuro-fascination.

Brain art can turn over its own ground of popularity if it chooses. The brain need not be a loud rock star dancing center stage. A Robert Ryman of brain art can emerge from the scientism of the realist laboratory and offer a snowstorm. Brain art can present the viewer with something terribly vague, "not ideational, much less securely optical," something where "Facture and the quality of paint qua paint loom large," as Hudson says when speaking of Ryman's all-white paintings.[2] Brain art could ultimately side with Kandinsky when he says,

> This world is too far above us for its harmony to touch our souls. A great silence, like an impenetrable wall, shrouds its life from our understanding. White, therefore has this harmony of silence … It is not a dead silence, but one pregnant with possibilities.[3]

Perhaps this is why lab coats are white, why lab rooms are white, and why fMRI machines are painted white. There is a recognition, inherently, in the sciences that the effort to grasp "harmony," as Kandinsky puts it, stares at a wall searching for possibilities. That is exactly what museum-goers love to do.

Brain art as shift

The museum brain, a kind of living post-mortem, acts as an anesthetic for our decline. But it is not enough to point back to existentialism's dread when high-lighting how neuroscience functions to alleviate the unsettled feeling that we, as humans, are materially unable to make perfect sense of our existence, including of our complex experiences. To make only that assertion is to argue that neurosci-ence is but an opiate to human understanding of how we come together and live this way or that. Neuroscience is more than an opiate—it offers viable treatments for debilitating medical conditions and advances human care. But, equally, we must recognize that we do not wish for any science to circumscribe a mode of being for us. Indeed, when we look carefully at brain artworks, we walk away skeptical that any neuro-delineations create more equitable relations or more moral systems of governance. Seeing a big glowing brain labelled "Male" or "psychopath" or "ADD," we are not too optimistic about the prospects. Blobs of general brain regions described in vague ways like "disgust area" and "emotion area" further reduce optimism. Neuroscientific interpretations seem too often to be recycled, like there are only so many plot lines: addiction mechanisms play like the story of "overcoming the monster," working memory and enhancing the anterior cingulate cortex sound like a "rags to riches" story, neuro-enhancement and implantable brain prosthetics are like "the quest," mirroring brain systems and all the high claims associated become like "the wedding comedy" because every-one suddenly throws a big, expensive wedding, but it all falls apart in a thunderstorm; yet, in the end, lessons are learned, the love is saved, and the couple gets hitched at the family's backyard BBQ. The others could simply be "tragedy," which is made possible first from a viable advance and second from a massive disappointment that strikes right at the heart of the hope of the advancement. Then again, the others might be the story of "rebirth"—but deciding on what they (re)birth is going to be context-dependent.[4]

Dampening expectations for neuroscience's ability to transcend social and material limitations circles us back around to face Chapters 2 and 3. Looking closely, we notice the first inklings of a return to the body already drying in the paint of the neuro-centric wall. In the chapter on neurosensuality, for example, artworks offer a brain of intimate spaces and of the coziness of home, anticipating a new role for embodiment and self-reflection on embodiment; that move itself anticipates a social and cultural shift, another way of living that changes The Brain in concept. Affective Realism, likewise, even when being composed as a brain of hard and untouchable facts, anticipates the viewer's doubt, which is also a shift toward another concept.

If neurosensuality performs a person being put at ease through neuroscience, then Affective Realism demands compliance by cracking a techno-scientific ruler across the knuckles. Taking these together, we feel a little weird. A little unsettled. So we shift. That is to say, brain art goads the viewer to make a turn toward the body and the body social. Feeling is felt as fickle. Environments are made evident as intimately affecting. Knowledge is witnessed as wrapped in its codependence on innumerable factors. As time lapses, things fall apart.[5] There is no The Brain to which we are beholden, brain art announces, unless we choose it that way. And if we do, then we perform a blindness to what The Brain itself anticipates.

The silent moment: a Deleuzian brain

To consider the radical relational philosophy of Deleuze at this point is to embrace unsettling change at a moment's notice. As scholars of Deleuze surely know, to encounter his work joyfully, one must dismiss a secure evolution yet be hopeful in continual recombination. With Deleuze, and this gets to the point, we pay greater care to how things shift beneath our feet.

With Deleuze, we can put The Brain in the mouth, chew it up, swallow, and digest it in the body. We then think anew about pathogens and toxic atmospheres. We hit the toilet and, in that moment of release, change irrevocably by turning toward an outside previously ignored or dismissed as unimportant even though it was obviously powerful. We confront the unmissable now in the bathroom. Turning face-to-face to stare into our own upheaval—a recognition of the force of the acrid taste of that factory apple—becomes an alteration and a reflexive state.

The stopping or the chewing, the eating and the drinking, the thinking or the thinking as a digesting, ushers us toward Deleuze.

At several points throughout this book, I have turned to Nietzsche to give more shape to The Brain, but I have not addressed Deleuze's readings of Nietzsche, despite Deleuze being foundational to many new ontologies foregrounding porous bodies and emergences with the environment. Thus, I want, here at the end, to think through what Deleuze has to say about Nietzsche's Will to Power. And I want to do so because Nietzsche's Will to Power encourages a glowing golden future, while Deleuze imagines our future to be extended and transforming both with us as well as far beyond us.

Here, the reader can intuit the underlying relation to the brain sciences and the competing views of the brain in works brain art. For Deleuze, the accessible relations of Things are not discrete, not always easy to locate, and those made actionable are only a small slice of what could be and what has already been passing by. So there's a strange intractability and unpredictability inherent in the conception of the world. Thinking between Deleuze's and Nietzsche's visions, I aim to re-see them as two ways of imagining the brain and, therefore, as able to elucidate possibilities for brain art. But first things first.

My broad view of Deleuze's discussion of Nietzsche is that Deleuze uses Nietzsche's character Zarathustra's and his curious encounter with the ever-changing god Dionysus—as detailed in *Thus Spoke Zarathustra*—as a way to support his own intervention in Western philosophy through privileging Becoming over Being. But I want to slightly complicate the meaning of Dionysus and diverge from Delueze's interpretation.[6] I want to say that the continual controversion of Dionysus, as a god, is really what Nietzsche offers to us, not Dionysus as a god to model ourselves after.[7] This is, in some way, hyper-Deleuzian. Let me put it another way.

Nietzsche, in my understanding, positions the fleeting present as the only place where we are able to release ourselves from a strong-headed pursuit of an Absolute refusing to change its directionality; this remains the case even when we adopt the idea of ever-Becoming. Noticing the peculiarities of the moment is the only place one can refuse to follow an Absolute; more so, moment to moment, one must refuse an Absolute (let me say "Absolute knowing and comfort" here to foreground a vision that elevates humility, questioning, and a remaking of the Self within the affective discomfort of continually encountering strange worlds). Following after Nietzsche, Deleuze reminds us to watch out for Absolutes, to seek to live in imper- manence and penetrability, but he defends the notion of Becoming Absolutely. To take the next step and contravene this Absolute, as Deleuze himself says that we must, we require openness to Becoming's controversion.

In view of The Brain, as detailed in this book, this *controversion* of Becoming (an anti-thesis) would be equivalent to the moment where we know absolutely that neuroscience will describe every contour of human and animal experience by peering into the brain. Obviously, to assert this controversion would feel like a reversal in the trajectory of this book, which privileges material contingency and sociality, generally aiming to make The Brain, yes, the brain again, but actually no, to make it the organism, no, the peculiar. But that, exactly, is the point and the glory of brain art, namely, the ability to get us to imagine beyond our dedications, to conjure the unexpected and the surprise, then spin us, dizzy, back around to an unknown future.

In the next few pages, I argue that the image of Dionysus in Nietzsche's *Thus Spoke Zarathustra* may not be important because of Dionysus's congruence with the idea that Being is Becoming, as Deleuze suggests.[8] Rather, Dionysus can just as easily be understood in the narrative as a trickster who creates a type exactly like himself and, thus, stands as a repeatable, which controverts his own Being as Becoming. Although he rejects stable Being by proposing the inevitability of Becoming, he gives back stability to Becoming in asserting it as a finality. This is signaled in a subsequent pause, or silence, when speaking to Zarathustra about himself. The move positions Dionysus not only as something inarticulate in advance of itself (in advance of Becoming) but also functions as a failure of Being as Becoming to live up to its promise to not compel another to enter its own ontological space and condition, i.e. a failure to be something other than another domineering god. As it turns out, Dionysus is a god after all—and one the same as the past ones.

This reading, I believe, goes one step further than Deleuze by not presuming, as he does, that coming to grips with the image of Dionysus is a resolution to a problem that philosophers after Nietzsche highlight, namely, the problem of rooting one's self in a stability when living in constant changeability. Reading Nietzsche again I, instead, suggest that there is no clarity on any resolution. But there is a rest offered. This "rest" resides within the lone moment where we, as readers, recognize what it means that Dionysus has just attempted to delay an answer to the question that Zarathustra asks.

All of this requires some discussion of the text, of course. But I should say up-front that I recognize that my reading risks adding more than what Nietzsche intended. But it nevertheless aims to emphasize the value of embedding oppositions and multiplicities into intellectual trajectories. And it complements creative approaches to brain art in so far as material engagements productively make multiple performative realities visible to us. The constant revision of readings and alternative views is inherent to the works that have these availability within their material. This is a kind of hermeneutics of revision and suspicion shaped into an ontological inquiry, and adopting it, we walk off sensing that material worlds are largely unknown, much deeper and more diverse than ever yet imagined, going further than the concept of the performative, themselves controverting any present Becoming, stretching into the insensible.

Scan 16: The Brain can be Other than another god

To discover intimacy with the uncertainty of materiality means desiring blow-your-mind mystery, a world where Things live in a million ways. Ontology is underdetermined. Bodies are alien domains. We must allow the wet ink of our ontological inscriptions to bleed across the page a little.

Once abandoning metaphysics, materiality strides center stage and announces itself as the new conspiracy theory. But just like the unreachable ideal world of Plato, we seem also to be locked out of the many worlds of Things too.[9] Answers to the state of Things are too complex and/or secretive for us. As Thomas Nagel now famously asked, "What's it like to be a bat?" For him, failure of human imagination coupled with inaccessibility to the subjective experience of other animals results in the conclusion that there are "facts beyond the reach of human concepts."[10] Knowing what it is to be a bat, or any other thing, is an approximation and speculation.

Prefiguring the more general lack of certainty about the world, Nietzsche composes as first principle the staunch declaration that "nothing more remains to which man can cling and by which he can orient himself."[11] Nietzsche pulls no punches. The structuring force of history and of identity collapses for him under the weight of that (colonial and oppressive) history and the subsequent revision of identities. Thus, Nietzsche declares, "God is dead."[12] So what now? He [meaning Nietzsche] pushes us today to grapple with that statement and form some kind of response.

For Nietzsche, the answer is to psychologically overcome the Will of Man. To live as an "overman" (The Übermensch). This means creating a new Will aware of the oppressions of Modern Man and the Self forged from enculturated dominations, which in Nietzsche's time seemed to be mostly religious dominations.[13] The point is: the eradication of a stable Self and a guiding force requires a revised way of being, thinking, and doing, which Nietzsche frames as a kind of motive toward the future, namely, to escape what feels oppressive and to embrace life in a more vibrant but raw or given form.

The question today, perhaps as ever, is whether people can be so flexible. Who can become aware of the present enculturated dominations, but more to the point, can this person then consider the abandonment of a guiding force when being inscribed as the medical subject and the consumer and, of course, The Brain? What happens when one organ of the human body is crowned The Lord God Almighty? If religious structures in contemporary society (for lack of influence or salience) cannot so easily be propped up as the prime antagonist anymore, then Nietzsche's statement "God is dead" reads, rather, as something broader—a social intervention screaming, "we need not be like that at all unless we choose!" It is a philosophical innovation aiming to generate a more free form of existence.

Nietzsche did not narrate the overman while being unaware of an obsession with re-creation. He seemed to know that the resolutely dead would endlessly rise again and has a distaste for history's revisionary recurrences. As Deleuze says when speaking about Nietzsche's Will to Power: "Why would man have killed God, if not to take his still warm seat?"[14] In like manner, Heidegger notes that the lack of direction that Nietzsche proposes instigates the search for a new composer: the very abandonment of God foretells a sociological and psychological replacement. But big shoes require big feet.

Something powerful enough to withstand the awful despair of not knowing much of anything about the universe, the ridiculous uncertainty of tomorrow's being, or the weird inability to explain something as simple as the beauty of touching the fuzzy leaf of the Stachys byzantina is needed. Upon finding a suitable replacement, and then another, new values and ideals inevitably arise as arbitrators of fact. But these arbitrators, as all do, slice off other possibilities.[15] What happens if The Brain shifts into the power position? Can we declare, in triumphant Nietzschean mode, "The Brain is dead?" We now stand at the moment of its coronation. We may not yet be able to contemplate the possibility of its replacement, as we are already interpellated as a type of Being susceptible to its Will to Power.

But maybe The Brain will prove to be good at human governance. Maybe it will become Socrates' philosopher King, i.e. a trustworthy way to assert ideals, to inscribe Truth, and to free people from self-deceptions and false appearances. Then again, as Folse notes, the conclusion of a philosopher as a perfect King "would naturally be felt as paradoxical by most of Socrates' listeners because philosophers were perceived as people with 'their heads in the clouds' and consequently as manifestly unfitted for the realities of the political world."[16] But nobody sees neuroscientists or data scientists in this way.

That gets at a point: the very idea that a "lover of wisdom" should be in charge and that people need guidance from a place where learning is systematic and where dumb opinions can be exposed through technical assurances cannot be brushed aside. The need to establish a foundation for knowledge, to avoid counterfeit beliefs, and to build ethical statutes that enable thriving communities feels as desperate now as it did in ancient Greece.[17] This is why input into a computational equation must be declared as automatic and, thus, composed as uncontroversial; if the machine produces the image, then the image functions in a photographic way, that is, in a seemingly transparent kind of way independent of human bias.[18] And we are continually inventing technical strategies for securing against protest. Of course, the whole project looks a lot like a brain transformed into The Brain, i.e. a declaration of The Real so that mere appearances, weak opinions, and unfit imaginations can be separated and dealt with accordingly. But wait. Isn't that exactly the domineering project that Nietzsche suggests collapses under its own historical weight?

The Brain can be other than another god. Instead of a domineering force, it can become half of a dialogue in a conversation stretching across the polis. It can be an indispensable character without offering an absolute promise. It can be tied together with a broad love for a whole range of expressions of humanity. It can, to start, be simply a brain.

As Socrates says, the lover of wisdom does not "show a love for one part of it and not for another, but must cherish all of it."[19] We can extend the idea and say that the Christian call of Jesus to "Love one another" performs a reversal in so far as radical love of one's neighbor upends the dominating cultural force of Christianity. Love performs a continual re-visitation of humility because it demands listening to all and embracing them within their range, without imposing grids of determination. There is wisdom in not foreclosing an outside. Indeed, if Christianity, as faith message of love, controverts Christianity as human kingdom controlling others, then we might learn from this controversion: wisdom loves the whole, indeed all the universe, including the unknown parts.

In an era of ecological death and increasing biotechnological complexity, we must cherish a multiplicity. In so doing, we must look for all that appears, not forgetting the many alternatives and contradictions. Like any strong, internally secured knowledge formation—and especially ones geared to be managerial and governmental—we must pursue neuroscience within an understanding of what it Wills while cherishing the contravening of its Will. In like manner, we cannot ignore how we might be wrong. This is an epistemological politics of the outside. If we journey through the desert and choose to stick to one direction, then we go that way to see what appears to us, but we must do so without declaring that we have seen it all nor that all that we have seen were mere appearances.

When introducing Nietzsche's text, Deleuze first reminds us that Plato, in the dialogues of Socrates, leads inquisitors old and young to a final question: "what is it?" Deleuze takes the question as a mode of thinking, one harping on essence and origins. He highlights "the distinction, dear to Plato, between beautiful things" and "nothing but beautiful." In this, Deleuze asserts that "it is undoubtedly a blunder to cite something beautiful when you are asked, 'What is beauty?'" Yet, for Deleuze it is also a blunder to ask about essence.[20] Here is where Deleuze finds Nietzsche so inspiring. In making Dionysus the figure that gives Zarathustra ontological insight and direction, Nietzsche fundamentally foregrounds plurality and rejects the old metaphysical tradition. The moment occurs when Dionysus suddenly interrupts Zarathustra's question about essence with palpable annoyance, saying, "'—which one is it? You ought to ask!' Thus spoke Dionysus and then kept quiet in his own special way, that is to say, in an enticing way."[21] The idea is that the more appropriate answer to a question of essence is yet another question: "which one?" Deleuze reads the passage this way:

> Dionysus is the god who hides and reveals himself … Dionysus is the god of transformations, the unity of multiplicity, the unity that is multiplicity, the unity that affirms multiplicity and is affirmed of it. "Which one is it?"— It is always him. This is why Dionysus keeps tantalizingly quiet: to gain time to hide himself, to take another form and to change forces.[22]

Deleuze's interpretation highlights the primacy of a Will to Power in so far as Dionysus can be seen as the figure of the Will to Power. "We should not ask 'which one wills?', 'which one interprets?', 'which one evaluates?', for everywhere and always the will to power is *the one that*."[23] Put another way, asking "which one?" is a rhetorical strategy to notice that any Thing is always composed as it is because of its enduring difference in a reactive mode with the environment. Thus, every Thing is a Will (to itself, to its direction or drive, its Power). Likewise, the Will of any Thing is itself always changing in its Becoming. Will is "a differential element" and a transmutation, something "inseparable from each case in which it is determined," a condition of forces, not an anthropomorphism.[24] For this reason, Dionysus, as a Will to Power personified, delays more speech. Or, the Deleuzian reading here seems to be that Dionysus signifies the act of difference, a shift or change in expression of Will by no longer speaking after confronting Zarathustra. He is but being himself.[25]

However, if we play with Nietzsche's text at this point to see what else it can Become, then we can imagine that Dionysus' delay to speak anything more is not an instant needed for a metamorphosis, as Deleuze suggests.[26] We might, rather, imagine that Dionysus does not know the answer to Zarathustra's question about essence. That is, Dionysus, in Being an unstable relationality, has no self-awareness of what sustains or drives him at any moment, so he, therefore, is unable to say anything to Zarathustra.

In this case, Dionysus' deferment after stating "Which one is it?" might not be interpreted as a call to see Being as Becoming or transformation, but as a kind of trick—to make Zarathustra think that there is access to knowledge about the "state" of all Things (Becoming) when Dionysus knows that there is no means of access (or said differently, when Dionysus does not know anything actually). We can imagine that Dionysus lies or represses the notion that he knows not how nor whether he is Becoming or not, even though he seems to be in a process of Becoming, and even though Deleuze reads Dionysus' response as an authentically delivered message.

Or perhaps Dionysus creates the subsequent pause not (only) to give himself the time to walk with the forces of Will to reject representation—all that is static and obsessed with the Thing—but also to impose his Will onto Zarathustra, to make Zarathustra more like him, i.e. to make Zarathustra hopelessly unstable, and in so being, abandoned by all definition. That is, to exercise a Will, as Dionysus must, he aims to make Zarathustra a type that is a differential from Essence, i.e. a wild plurality. But, as I have said, when Dionysus asserts Being as Becoming, he has no idea how or why he Becomes; he has no answer for himself in being a difference or a shift, so he pauses (to shift from speaking). Yet, in this act, he also in some basic way related to "the pause" controverts his own Becoming. Here I suggest a more salacious interpretation for the reader: that Being as Becoming is, itself, a form of philosophy seeking a repeatable. Indeed, the only way to do anything more is to allow Zarathustra to fill in the gaps, that is, to speak it again for himself, to re-animate Becoming. To be a second and a return, a kind of repeatable or representation of Becoming.

Deleuze suggests that what Dionysus wants is what a Will wants:

> What a will wants is not an object but *a type*, the type of the one that speaks, of the one that thinks, that acts, that does not act, that reacts, etc. A type can only be defined by determining what the Will wants in the exemplars of this type.[27]

In saying this, Deleuze suggests that Dionysus, as a character in the book, prefigures Nietzsche's methods. We can say that Nietzsche's method is discovering trajectories of Becoming in the figuring of types. But the only type in the silence that Dionysus leaves after his deferral to another question is an old type, which is an accessibility to a form of Being, an entrance into that form. That accessibility and entrance is the instigation to understand Being as Becoming, Absolutely, regardless of what Zarathustra does or says.

Here is where my own argument takes its full shape. If the rebuke that Dionysus makes before Zarathustra is a presentation of a type "of the one that speaks," then the rebuke is a *should* and, thus, a domination. Further, the broken empty hanging air that follows the rebuke, a space of nothingness, is also a type for Zarathustra to Be/come. That type is not a linguistic expression but a performative one; it not only opens space for Zarathustra to Be-come, but it performs the nothingness that Dionysus desires Zarathustra to Become. However, the type wanting another like itself controverts the first, namely, Becoming. In that space of silence, Dionysus' Will controverts itself, even if it is a differential reaction—because silence requires Zarathustra to take up the same type as Dionysus. To fill in the gap or be "an again."

Following this line of thought, we might also argue that there is no "method adequate" that traces differentiation; we can say this if Nietzsche, as Deleuze argues, intends Dionysus' deferral to another question as some description of existence.[28] How could "a differential, typological, and genealogical method" hope to capture Dionysus' slick changeability or his sly silence?[29] Tracing the formation of any type seems caught up in the question, "which One?" But more so, how much harder is it to trace a Becoming that must defer to silence to account for itself, i.e. cannot know its trajectory or comprehend its own relevance? Any charting of a type is, of course, only possible in interpretation—making "an again" from a representation.

Moreover, what do we make of Nietzsche's claim that "God is Dead?" For Nietzsche, the God denounced is not only the Judeo-Christian God or moralizations springing from the metaphysical drive to answer "what is it?" but any god. Thus, we should include Dionysus if we are going to take Nietzsche at his word. Especially Dionysus. Anything that answers with an affirmation (of what *should be* asked) is never honest in saying "which one?" Here I read Dionysus again as a trickster and liar. He is a distraction to his own philosophy of life, even as he is Becoming. He is one and the same as the ultimate represented.

To the point: in asking a question to an asked question ("which one?" to "what is it?"), we see another ontology offered that is but another error or a falsity. Dionysus is an ever repeatable. His answer, "which one?" shows a Will to Power as infinite—ever returning as the condition of Being as Becoming. Ever explanatory even if having no defined explanation. In this general evaluation, I agree with Deleuze. But I cannot find a total rejection of metaphysics in Dionysus.[30] I do not take Dionysus as showing the reader an "adequate method," but merely offering another flat rejection, or controversion, of the statement "God is dead," precisely by offering an eternal condition. In Dionysus, God lives. Life cannot go on otherwise. This re-inscription is another form of Being even though it is composed with a concept of Becoming. It asserts no transcendence from its own concept and does so, perhaps more distressingly, by making us fill in the gap ourselves, "tricking" us into a defined form through the promise of renewal.

Of course, Dionysus is also a contradiction if lacking knowledge about what forces sustain him. In that instance, we guess that he does not know if he is Becoming or Being or if he is always One way (doomed to a life of ignorance) or could ever be any different. He is only able to do what he does, which is to defer the answer in pointing elsewhere. Dionysus is like a man about to be arrested for a crime that he knows nothing about, unsure whether he committed it or not, and who quickly points to the sky and says, "Look" while he runs away from the police.

Within this alternative interpretation, there is no rational, solid answer for Zarathustra, except in realizing that there is no "good answer" to any of his deepest ontological questions. There is no escape from questioning, except in silence. This is the better lesson that Dionysus teaches.

If there is any celebration in Dionysus as demonstrating for the reader a Will to Power, then it is not in interpreting his answer as rightly banishing metaphysics in favor of Being as Becoming. That would be another dominating objectivity, another affirmation, even in being a paradox. But the freedom celebration, or perhaps the consolation, is the suspension of a moment caught in a Being or in a Becoming, or whatever we can think at a moment. This is the sharing of a secret, a secret that holds no representation, supports no image, constructs no discourse. What then is this silence but a stillness, and what is that but itself?

There is a rest in the recognition of the vast universe so unimaginably varied, an epic sea, a lyrical geometry whose twists are impenetrable, terrifyingly recursive and opaque, an endless dawn of stupidity, a majestic dream inside of the rippling red and black threads of a block of marble, traveling on without end. Stillness succumbs to the embrace of the incomprehensible universe. It is a type opposed in every possible way to domineering. The stillness is union, and that union is an acceptance. That acceptance is a love.

What does it love? Perhaps most of all, it loves an expanse. It loves the possibility of another to such an extent that nobody, not even Deleuze or Nietzsche, determines whether it will Become anything more or ever Be radically different. To say then that Deleuze or Nietzsche are good or right philosophers by offering us Dionysus is to demand a controversion of that claim. The philosophical embrace of the expanse cannot but dedicate life to desperate humility, just as the insurrection of that embrace will eventually embrace it once again.

Scan 17: there is no brain of any kind in brain art

We can, if we choose, revise how we see this thing called the brain.

When neurology is made into an art practice, neurons are not brainy things; they are not of or inside the brain at all. They are paint strokes, light machines, metal stands, hands and feet. Nothing looks the same anymore. Machines, oxygen tanks, surgical masks, skulls, blue blobs, black spots, feelings about Mom, all are given new life by the strange places where brains are encountered as material reinventions.

In being recomposed, there is no brain. In those moments of eruption and reassessment, a line of ink makes something else. The artwork whispers: "there is no brain here, only a materialism and your muddle of moments and compacted histories."

In brain art, as in neuroscience, the brain is a representation (René Magritte grins). Brain art extinguishes the brain as suddenly as The Brain. This is the case even when brain art props them up to put them out on display. The gesture, the irony, must be recognized for brain art to succeed; it returns The Brain, as ideal, back to representation, through which it came to be idealized.

The first success of brain art, then, is the constructivist vision applied to the practices of neuroscience. The brain organ comes into appearance as a kind of brain—a formation predictive of The Brain—composed through strategic practices. But the second success is the commentary on representation itself as overwriting the vibrancy of materiality. Photographic allure and psychological trickery flatten and simplify. Representation puffs a laughable remark: the 2-D, time-stamped paper depiction corresponds directly to Nature. But in our strange material encounters—carving a brain from a bar of soap, for example—the artist gives agency back to materiality, allowing it to reassert itself; materiality then reintroduces doubt about a bifurcated world—the real and the representational—haranguing the notion that they live together in harmony. They never were in love. Or the love was one-sided. Perhaps they never even met.

Intimately, for us, when seeing brain art, we have an opportunity: to realize that brains, once inscribed, are technological reductionisms and symbolic punctuations. In brain art, we have the means to realize that adopting neuro-discourses or applying concepts of normality and abnormality to ourselves and our families is to live within the scoped image of another, indeed whole sets of Others, including study participants, scientists, lab directors, grant funders, journal editors, colleagues living and dead, fMRI machines, computer software programs, software designers, technology business managers, the list goes on. We may be transfixed by the scientific review board, but brain art, like other great resistances, opposes anything being converted into an us.

The joy of living without a brain

In *The Joy of Sex*, Alex Comfort opens the first chapter by saying, "Love, like singing, is something to be taken spontaneously."[31] The joy of singing is belting it all out in the kitchen, feeling the song run through the body, letting it take control. Try singing Mina's cover version of *Se Telefonando* in all three keys as loud as you can. The spontaneity of that joy, as with a fulfilling sex life, requires an openness to the moment. But that moment is not an exercise of individual desire but a moment mutually created. Comfort argues that "adventurous and uninhibited lovers" attune to each other. They are not "obsessed with one sexual trick at the exclusion of all others."[32] The joy of sex in spontaneity means that mutual openness and acceptance leads lovers into wild nights or slow, day-long rendezvous: together they move moment to moment.

I return to the body to contravene Freudian scopophilia, or the desire to fully expose one's self and others,[33] a desire that I read as swimming somewhere within the visual drive to make neuroscience a totalizing, penetrating technical gaze. The voyeuristic activity that Freud argues as an inherent sexual desire—emerging around the time when children learn of toilet functions and strive to witness the forbidden—can be hyper-realized as an abnormal looking, which gets restricted to the genitals and transforms into an obsession that supersedes sex and, ultimately, gives rise to disgust. Freud details this as a form of perversion.[34] And it is easy to see how scopophilia is a self-obsession, a singular drive for a singular individual without regard for the Other. We can also see how the perversion is uninterested in the present moment of nakedness despite desiring it. That perversion is the need to repeat the act, a longing for yet another presence, even amid total nakedness. We might link this longing to an inherent ontological absence being felt, our own inability to see and understand any Thing in its entirety, as whole. That absence, a terrible awareness (a presence) of the absence of sight and of knowing "all the angles," so to speak, is also an inability to sight the Other. Accordingly, it is an inability to fully care for the Other (having so many unknown and dynamic aspects). The consequence is a desire to seek after ever-greater exposure, which operates like a coping mechanism. But the need to repeat the act results in the loss of the given moment. In missing the moment through desiring yet another, the sight of the Other as naked but not ever in-presence exposes oneself as incomplete.

Controvening scopophilia requires an embrace of the moment joined to satisfaction with partiality—a joy and a love for the hereness, i.e. an event, an appearance, a moment nothing more or less than whatever it is. Such alertness and particularity makes scopophilia impossible.

To contravene Freudian scopophilia is to embrace what nakedness presents as itself in its moment. To overturn Freud is to transform the ontological lack that repeated exposures seek to overcome by resting in the moment, taking it in and for itself. Likewise, to overturn ontological domination, we can choose to be suspended between formation (a body and an arrangement) and non-Being (a not yet, a next emerging), satisfied with both appearance and dis/unappearance. The world is an opportune moment for the living.[35]

But to simply assert the need for a feel-good "embrace" of any presence leaves us suspended. Do you feel it? … We must move again …

To step out, we anticipate; what do we anticipate? Spontaneity. Turning back to neuroscience as an example, we can see the image, the activity of the brain, the structure of the experiment, the insight of the scientist, each as a moment of joy. Those moments are joyful precisely because they are not fully transparent. They highlight a suspension of the living arrangement. They offer a brain, and then the mind opens to that brain. In those moments, we feel the mutual spontaneity. We fall in love with what we have because in having something, we have what was not a minute ago; we have what could never have been fully expected. That is, after all, the point of experimenting, is it not?

The joy in the exercise of the experiment, counter to how we might imagine the scientific enterprise, is in *not grasping resolutely* any One, not relying on One finality, One adoption, not declaring One Brain. The joy is the moment that is—because of itself—something more to live out and to do. But it is also already more than that—it is the immediate wow of what else appears and can appear when the moment appears. It is the previously unimaginable resonance and the implication that has thus far escaped us.

I am speaking about something akin to meeting a lover for the first time knowing that it will take years to get anything more than a glimpse of the depth of this person. The excitement that a stunning first date generates and the joy felt then—the electric buzz of possibility for the relationship, which keeps you up late at night thinking about every detail of the dinner conversation and every facial expression. It is a celebration of itself as well as an exciting anticipation of another moment, but not One, not One lover, not One ending. This joy recurs through a long relationship, and when a new flash of personality or a new story rushes out, it surprises and delights because of the returning realization that one's own partner is an immensity that lives despite presumably being so well defined. In discovering our ignorance, we feel joy.

Thinking with another kind of brain, a brain unable to be delimited, we can witness with joy the domineering The Brain functioning as a technê because it forces us to move again. We try to understand our own inner life through the popular conception of The Brain, and we accept with joy that it must then lead to yet another brain. We see before us an altered image of ourselves. And we feel joy because The Brain already embodies the death inherent to the mythical version of neuroscience being bantered about as a primary reference point for the human. And we are happy that the scan is pretty because it might be really ugly one day. And here, outside of itself, escaping its own scaffolded security, The Brain can be seen in the calm of silence. We no longer need to hope for neuroscience to finally save us or to lead us to a neural transcendence. We can accept that it will do this and also will not do this at all, not of its own. We anticipate a completely unexpected brain from a different point of view: the human with an instantaneous and unpredictable Being, i.e. the human without The Brain.

This is the joy of living—to move with but also beyond inscription. Whether we see ourselves moving in ontologically entwined and transformative relations or not, we move in an unsteady way, which may or may not prolong our life, confirm our prejudices, or give us better days. The inability to control extends to each moment. So if we direct ourselves toward some future that we imagine now as an improvement, we can discover a tomorrow that upends the whole project, and in the turmoil, we can embrace the joy of that spontaneity.

There is joy in brain art. The idea that "There is no brain" captures this sensibility, as does "The Brain is dead."[36] Clearly, we have an organ that we like to call the brain, but just as clearly, we have no organ disarticulated from the world and able to fulfil the promises of Capitalist technoscientific endeavors. Brain art makes this quite evident. And the tension between these—a brain and no brain, just as with The Brain and a brain—performs for us an absurdist play, and we find much joy in watching it unfold.

"The drive to order" sits side by side in a French café with an artist named "Inevitable alteration."[37] They discuss the concept of predictability. The artist suddenly says that she feels quite hot. In contrast, the drive to order knows that she has felt quite cold the entire time. They take another sip of coffee, which one presumes to be too strong while the other presumes too weak. The drive to order says, "Look at the shape of the sky, think about gravity!" The artist responds, "What if it were different, and what would you say then?" They both laugh, although neither is sure why the other is laughing. They both presume that they are right. But they both sense that somewhere they have been in the wrong. So they smile at each other. In that moment, they feel a wisp of joy. The artist reflects, thinking that this joy comes from the idea that nothing really can be ordered after all, and the surprise of her unexpected laugh and sudden smile is the joy of being always right about alteration—but wait, does this mean the artist is wrong? No matter, she thinks, I will just change my mind anyway. Just then, the drive to order thinks that the joy comes from seeing just how the artist will always disagree and, therefore, all things truly do sustain an order; the joy comes from recognizing that their disagreement, in fact, falls out from their orders. Still, they can share a laugh and a smile. How wonderful! This is the point in the absurdist play where we turn on ourselves to laugh, where we see their inherent contradictions, where we apply those contradictions to ourselves, and in that, we discover joy.

Scan 18: the brain can be reimagined

Brain art is a vehicle for rethinking our attractions, for reassessing our realisms, and for illuminating our feelings about how the body is explained to us. In this, brain art participates in a broader intellectual effort to "traverse the fluxes of matter and mind, body and soul, nature and culture."[38] Many works discussed in this book imply that biology and sociality need to be further combined; they re-philosophize science and simultaneously re-scientize philosophy. They speculate on making matter matter differently.

Of course, not all brain art proves revolutionary, even if it sometimes upsets and disrupts. Like much of the techno-body art that media theorist Jaonna Zylinska examines, the brain art examined in this book can appear "too deterministic, too prone to techno-hype, and uncritically fascinated with the technological process itself."[39] However, brain art, often simply by being made, advocates a sensual approach to bodily materiality. Attention to biomedical experiences or feelings of intrusion or fears of an unwanted diagnosis allows those experiencing the artworks to ponder how neuroscience might revise claims or, conversely, flourish when put in conversation with embodied responses to it. Brain art disrupts objectivity in tangibility just as it expands the view of what is untouched.

The material confabulations of brain art mesh phenomenological reflection with technical, representational investigation and end up splattering existential dilemmas all over the wall with a wet brush. Whether we believe that Things wildly come together through complex ever-shifting relationalities or that Objects like the brain take defined shape internally for and within themselves, we can walk away feeling uneasy about our position. The possibility for the closed hand being an opening and the open one being a closing makes a model intellectualism. Having another view when touching the soft quilted fabric of a frontal cortex, and then another when seeing its bold colors, both perhaps completely unexpected when wandering into a "brain art exhibit," makes for futures where biology and sociality cannot be so easily divided or defined.

If The Brain can ever be reimagined, then it will likely be through brain art. But artists will need to go further than showcasing an uncontroversial and beatific assemblage of neuroscience discoveries. Brain art might need to get stupendously unreal. It needs greater dispersal; slip over itself much more easily and fill the room with banana peels; it needs to function as a thought outside itself and create a peanut trail of firings in open air and then ask the audience to ingest them.

The joy of brain art

The joy of brain art is the creation of the unknown. This is another reversal.

The stock commentary about the joy of art, as many in the humanities as much as the neurosciences surely know by now, is in exposing something more, making something new through interdisciplinary alignment. The glory of brain art is the way that is it marketed as moving audiences into uncharted affective terrains, exposing as-yet undetected influences on mental life, manifesting subliminal connotations, and creating innovative knowledges. The usual celebration centers around hidden bodily realities formed when artists turn toward the sciences. The sciences, we suspect, should be thankful and learn from the sensual approach, while the arts, we suspect, should engage more seriously with "the actual" material Thing if spending so much energy contemplating bodies. However, I see the joy of brain art to be more powerfully composed as a release. The here and now of disciplinary knowledge can be battled out by opposing sides, but what brain art offers is a back-peddling, another re-evaluation.

Putting the tools of neuroscience to use in creative purposes builds a wildly unpredictable brain whose material formations are blended and interpretations are up for grabs. The lovely network of geometric intensity brings joy when the image overlaying charts and graphs remains recognizable as a call to critical interrogation. The joy is not neurobiological perfection or universalization but seeing the psychological longing for that perfection or universalization. This is a joy because brain art compels us to turn back to ourselves and relish a Being that in Becoming strives nevertheless to be minutely circumscribed. This is a comedy. We laugh because we love it. And we laugh because we do not know, for sure, do we, how much confidence we should have in our persistent correlations. We recognize ourselves in these moments of creative conglomerations, and we back-peddle a little bit on our absoluteness.

The joy of brain art is the absence of a brain in the artwork. The foregrounding of material mash-ups that look not equitable to a brain give a giggle. Highlighting the construction of the brain as "the brainy thing" makes some fun of organs and cherishes a little mayhem at the formation of a brain (how it works and what parts do what); brain art makes the brain disappear across the innumerable delineations of neuroscience findings spread out over years of study and thousands of experiments and interpretations, dis-locatable and untouchable as such. The viewer can wonder where it all went. And then we might speculate: How could we make a more viable representation, more exacting, more "me"? Then the answer slips off the edge of the frame. So many molecules, so much material nuance, so many subjective experiences, the artist reminds us. So much time. So much interchange. So much selection, deflection, reflection, and decision around resources and stimuli, input and output. Yet there is joy: the experience of the exhibit brings a self-satisfying joy in thinking that no one could inscribe the molten human immensity of being an "I" on the museum wall.

The intrusion of the brain scan trying to mark a moment of neuronal life down to a 30-inch image on a 24x36 monitor brings a double joy. The first is in seeing that the phenomenal "I" remains far beyond technoscientific surveillance. Escaping the limits imposed by the fifteen undergraduates used in the study is the joy of thinking past the overlay. Brain art pushes our minds to wander far beyond the computational conglomeration to a place where we are not equated to machines. We fly past a planet where brains do not "code" information. The joy of brain art is the total loss of appearances, like the high school bully breaking down in tears when watching *Free Willy* for the first time. The joy of brain art, as with the brain scan, is being able to see what escapes sight in a sudden and obvious way. The brain detailed in technical precision like finely cut glass offers, of course, the immediate pleasure of an aesthetic moment, but it also offers the joy of saying, "God what if we were all just like that? How boring! How despicable we would be!" This first joy, then, is the absence— something missing, something more—which is often more intimate to us than our presence.

This brings the possibility for yet another joy: the realization that we could also be so terribly wrong. Neuroscience may well progress with digital tools to a point, one day, when the most intimate affective and emotional expressions of living as "I" can be detailed and even experienced by others. But inside of this science fiction imaginary, we discover a new joy in thinking what it will feel like, in that moment, to share so fully in each other. How much joy we would feel!

In being open to reversals, there is no need to be so limited in our conception of what a body *is* or what it *means* to be a body. Brain art can, in its best performative and undulating modes, reinstate the joy of unknowing. And it can do so—this is how far joy can go—even when perpetuating the old adage "reason=virtue=happiness," even when propping up the statistical real as the neurobiological real, which translates so naturally into a claim about the psychological real. It is worth remembering that the love of neurobiology made as art can always break the bonds of the real with one small gesture—the insertion of a smiley face in the amygdala, the extra sticky stroke of black paint across the shape of the thalamus, the use of glitter for the BOLD signals, the Navy Seal as the prototypical neuroscientific test subject, the scan of the rat brain with "PSYCHOPATH" written over top. How much joy there can be in thinking that we can upend our deepest commitments with a stroke of paint!

If we choose, there is joy in a suddenly new way, an unforeseen, an unimaginable. There is joy in seeing where we were wildly wrong. How much there is to know! The thought can also bring much terror, but whether it brings terror or joy is not a matter of what is known at any moment. Rather, the determination is about how one approaches the unknown as an unavoidable.

In not avoiding the absences in presences, in loving the moment of the unknown, we love our selves without any precondition for anything more. Returning, then, to the Socratic equation that so bothers Nietzsche—"reason=virtue=happiness"—we might choose to rid ourselves of the idea that personal satisfaction is an outcome, i.e. "happiness," and then forget entirely about the isolation and solidity of the starting point, i.e. "reason," but rather work from the gut feeling that existence is an immeasurable expanse. What joy! From there, we can embrace equations, as we must do to live, even as we sit in the open field and feel a bit lucky to be there. From within our commitments, taken to be equally subject to surprise as ontological conceptions, as ultimately expungable in being anthropocentric constructs, we can breathe in the feeling of suspension. We might in that moment of silence consider whether we can turn back and go another way. We consider whether there is One underlying this suspension. There may well be after all.

The reliable math of set theory, Alain Badiou argues, is always more, more, more, always multiple.[40] In that mind-blowing equation, we discover an unreasonable immensity as the greatest controversion of the very reliable, repeatable mathematical reason that generates it and that it relies upon. The equation asserts a One that is not singular nor easy to grasp. It leads only to an expanse while yet still being defined and still yet being a controversion of a universe that cannot be in any way defined. If only The Brain could be like that, and if it were, then it would be unrecognizable.

Notes and References

1 For works by Chuck Close, see "Chuck Close," http://chuckclose.com/.
2 Suzanne Hudson, "Robert Ryman, retrospective," *Art Journal* 64, no. 3 (2005): 65.
3 Wassily Kandinski, *Concerning the Spiritual in Art*, translated by M. T. H. Sadler (New York: Dover Publications, 1977), 39.
4 Kasia Boddy, "Everything ever written boiled down to seven plots," *Telegraph*, November 21, 2004.
5 Things Fall Apart is obviously a reference to the book of that same title, by Chinua Achebe. See: Achebe, *Things Fall Apart* (New York: Penguin Books, 2007); thus, it is a reference to a Modernist view toward increasing darkness and chaos amid a struggle for identity and against colonization, which is herein applied to the hopeful optimistic view that we can outline every contour of experience by looking at a brain.
6 Gilles Deleuze, *Nietzsche and Philosophy* (New York: Columbia University Press, 1983).
7 By "controversion" I mean the turning around on or turning back against one's self concept.
8 Ibid., 76–77.
9 The Apostle Paul says "we see but a dim reflection" in I Corinthians 13:12; see Berean Study Bible, https://biblehub.com/bsb/1_corinthians/13.htm.
10 Thomas Nagel, "What's it like to be a bat?" *Philosophical Review* 83, no. 4 (1974): 441.
11 The quote comes from Martin Heidegger, *The Question Concerning Technology and Other Essays*, translated by William Lovitt (New York: Garland Publishing, 1977), 61. Note that the phrasing of "man" is from Heidegger's text.
12 Nietzsche says this in several places but to see one location with some discussion of the implication of the phrase, see Frederich Nietzsche, *Thus Spoke Zarathustra* (London: Penguin Classics, 1961), 41.
13 Ibid.
14 Gilles Deleuze, *Nietzsche and Philosophy*, translated by Hugh Tomlinson and Michael Hardt (New York: Columbia Classics, 1983), 151.
15 Heidegger, *The Question*, 61.
16 Henry Folse, "The paradox of the Philosopher King," Philosophy Courses (2001), Loyola University. Available at: http://people.loyno.edu/~folse/Philking.html. Note: italics in original.
17 Ibid.
18 This observation derives from Beaulieu who says something similar in: Anne Beaulieu, "Voxels in the brain, neuroscience, informatics and changing notions of objectivity," *Social Studies of Science* 31, no. 5 (2001): 635–680.
19 *The Republic of Plato, Vol. I* (Cambridge: Cambridge University Press, 1969), 474c.
20 Deleuze, *Nietzsche*, 76.
21 Ibid.
22 Ibid., 77.
23 Ibid.
24 Ibid., 84–85.
25 See ibid., 77–85.
26 See ibid., 77–80.
27 Ibid., 79, italics not in original.
28 Ibid.
29 Ibid.
30 See ibid., 79–81.
31 Alex Comfort, *The Joy of Sex* (New York: Simon & Schuster, 2003), 8.
32 Ibid., 9.
33 See *The Standard Edition of the Complete Psychological Works of Sigmund Freud (S.E.)*, Vols. 7 and 9, translated by James Strachey and Anna Freud (London: Hogarth Press, 1886–1889).

34 For discussion, see Peter Mahon, "Scopophilia," University of British Columbia (course blog). Available at: http://faculty.arts.ubc.ca/pmahon/scopophilia.html.
35 Aristotle describes Kairos as "an opportune moment." I inherently reference the idea but seek to apply it to all emergences, all forms that appear into Being. For discussion of Kairos, see James L. Kinneavy and Katherine R. Eskin, "Kairos in Aristotle's Rhetoric," *Written Communication* 17, no. 3 (2000): 432–444.
36 These phrases derive from my article; see David Gruber, "There is no brain: rethinking neuroscience through a nomadic ontology," *Body & Society*, April 2019, online first.
37 I am referencing here René Magritte's "This is not a pipe." For discussion, see Michel Foucault, *This is Not a Pipe: With Illustrations and Letters by René Magritte*, translated by James Darkness (Berkeley, CA: University of California Press, 1983).
38 Rick Dolphijn and Iris van der Tuin, *New Materialism: Interviews and Cartographies* (Ann Arbor, MI: Open Humanities Press, 2012), 88.
39 Joanna Zylinska, *Bioethics in the Age of New Media* (Cambridge, MA: MIT Press, 2009), 149.
40 Alain Badiou, *Being and Event*, translated by Oliver Feltham (London: Continuum, 2007), 3–5.

INDEX

9780367898199

The first of its kind, this book examines artistic representations of the brain after the rise of the contemporary neurosciences, examining the interplay of art and science and tackling some of the critical-cultural implications.

Weaving an MRI pattern onto a family quilt. Scanning the brain of a philosopher contemplating her own death and hanging it in a museum. Is this art or science or something in-between? What does it mean? How might we respond? In this ground-breaking new book, David R. Gruber explores the seductive and influential position of the neurosciences amid a growing interest in affect and materiality as manifest in artistic representations of the human brain. Contributing to debates surrounding the value and/or purpose of interdisciplinary engagement happening in the neuro-humanities, Gruber emphasizes the need for critical-cultural analysis within the field. Engaging with New Materialism and Affect Theory, the book provides a current and concrete example of the on-going shift away from constructivist lenses, arguing that the influence of relatively new neuroscience methods (EEG, MRI and fMRI) on the visual arts has not yet been fully realised. In fact, the very idea of a brain as it is seen and encountered today—or "The Brain," as Gruber calls it—remains in need of critical, wild and rebellious re-imagination.

Illuminating how artistic engagement with the brain is often sensual and suggestive even if rooted in objectivist impulses and tied to scientific realism, this book is ideal for scholars in Art, Media Studies, Sociology, and English departments, as well visual artists and anyone seriously engaging discourses of the brain.

David R. Gruber is an Assistant Professor in the Department of Media, Cognition and Communication at the University of Copenhagen. His research spans the rhetoric of science, body studies, and the public understanding of science. Much of his work focuses on the role of neuroscience in society and the reasons why brain findings can be so persuasive. He is co-editor of *The Routledge Handbook of Language and Science* with Lynda Walsh and creator of the neuro-humanities website, Neurohuman.com.

Cover image: *Critical Thinking*, 15" X 26", a print by David R. Gruber.

PSYCHOLOGY / NEUROSCIENCE / ART

Routledge
Taylor & Francis Group
www.routledge.com

an **informa** busir

ISBN 978-0-367-89819-9

9 780367 898199

Routledge titles are available as eBook editions in a range of digital formats